MR. LIQUID CRYSTAL

MR. LIQUID CRYSTAL

THE UNTOLD STORY OF HOW
JAMES L. FERGASON
INVENTED THE LIQUID CRYSTAL DISPLAY
& HELPED CREATE THE DIGITAL WORLD

Terri Fergason Neal and Marian Pierce

New Insights Press

Editor: Rick Benzel
Art Direction and Design: Susan Shankin & Associates
Cover Illustration: Daniel Baxter
Figures: Tim Kummerow

New Insights Press
An imprint of Over and Above Creative Group
Los Angeles, CA
www.overandabovecreative.com

Copyright © 2016 by Terri Fergason Neal and Marian Pierce. All rights reserved. This book contains material protected under International and Federal Copyright Laws and Treaties. No part of this publication may be reproduced, distributed, or transmitted in any form or by any means, including photocopying, recording, or other electronic or mechanical methods, without the prior written permission of the author, except in the case of brief quotations embodied in critical reviews and certain other noncommercial uses permitted by copyright law. For permission requests, please email the author at following address: info@mrliquidcrystal.com

ISBN: 978-0-9973357-7-4
Library of Congress Control Number: 2016912202
First edition. Printed in the United States of America

Publisher's Cataloging-In-Publication Data
(Prepared by The Donohue Group, Inc.)
Names: Neal, Terri Fergason. | Pierce, Marian.

    Title: Mr. Liquid Crystal : the untold story of how James L. Fergason invented the liquid crystal display & helped create the digital world / Terri Fergason Neal and Marian Pierce.

    Description: First edition. | Los Angeles, CA : New Insights Press, [2016] | Includes bibliographical references.

    Identifiers: LCCN 2016912202 | ISBN 978-0-9973357-7-4 | ISBN 978-0-9973357-8-1 (ebook)

    Subjects: LCSH: Fergason, James L., 1934-2008. | Liquid crystal displays--History--20th century. | Inventors--United States--History--20th century.

    Classification: LCC TK7872.L56 N43 2016 (print) | LCC TK7872.L56 (ebook) | DDC 621.3815/422--dc23

Visit MrLiquidCrystal.com and Facebook.com/Mr-Liquid-Crystal

*Well, I know who'll take the credit—
all the clever chaps that followed.*

*Came, a dozen men together—
never knew my desert-fears;*

*Tracked me by the camps I'd quitted,
used the water-holes I hollowed.*

*They'll go back and do the talking.
They'll be called the pioneers!*

—Rudyard Kipling, "The Explorer"

This book is dedicated to:
Andrew, Laura, Sydney, Tim, Max, Erin, Allison,
Kate, Lucy, Annemarie and Maggie
embrace life's challenges and follow your dreams

CONTENTS

INTRODUCTION	XIII
History of this Book	XV
1. FROM A SIMPLE BEGINNING	1
Maternal ancestors	1
Paternal ancestors	4
Life on the prairie	5
Jim's siblings	8
Jim's youth and interest in science	10
Meeting his future wife, Dora	14
College years	16
First professional job	18
2. A WESTINGHOUSE ENGINEER	20
The influence of Max Garbuny	23
First born daughter . . . and first liquid crystal experiments	27

3. A UNIVERSE OF DISCOVERY — 30

Digging deeper and discovering possibilities — 37

The chemistry of creating a room temperature liquid crystal — 38

Jim's chemical success — 45

4. THE WORLD'S 1st LIQUID CRYSTAL-BASED DISPLAY DEVICE — 49

Inventing the first cholesteric optical temperature measuring device — 51

5. INCUBATION OF THE LCD AT WESTINGHOUSE — 57

Jim's assistants at Westinghouse — 58

LCs color changing speed — 60

First liquid crystal display — 62

Early competitors in LC research — 68

Popular science magazines take notice — 70

Westinghouse problems — 71

The 1965 Liquid Crystal Conference and the move to Kent State — 73

6. THE LIQUID CRYSTAL INSTITUTE — 77

Getting LCI off the ground — 78

Developing business for LCI — 82

Life magazine makes liquid crystals, and Jim, famous — 87

7. GREAT FRIENDS	**92**
Alfred Saupe comes to KSU	97
Jim's extensive presentations	98
8. SUCCESSES AND CHALLENGES AT LCI: 1968	**100**
A rejected article and sour relations	108
LCI and the Vietnam War	111
The complete breakdown in relations with Glenn	114
9. THE LAUNCHING OF ILIXCO	**118**
The new company needs a name	121
Inventing the twisted nematic liquid crystal shutter	124
Time for investors	132
First customers	133
Advancing the understanding of LC's	136
Getting shut out of his position at LCI	137
10. THE PEANUT CAR EUROPEAN SOJOURN: JUNE–AUGUST, 1970	**142**
Learning new techniques in Europe	147
Back at LCI	151
11. ILIXCO'S DEVELOPMENT OF THE 1st TWISTED NEMATIC PROTOTYPES	**152**
The first TN patent	155
The beginning of the patent battle	157

No proof that either Helfrich or Schadt invented the TN-LCD	160
Going into production	161
Building the first TN-LCD watch	167

12. MAKING A MANUFACTURABLE WATCH DISPLAY — **175**

Growing the company	178
New solutions to TN-LCD challenges of surface alignment	180
Solving the reverse tilt problem	188

13. THE "TIME COMPUTER YOU WEAR ON YOUR WRIST" — **190**

Forming the display's glass envelope and sealer	195
Adhering the polarizers	200
Sales begin	201
The need for more finances and a contract with Gruen	203
Funny employee stories	204
And frightening stories, too	206

14. PATENT NUMBER 3,731,986 — **209**

The patent battle begins	210

15. HEARTBREAK AND ANXIETY AT ILIXCO: 1973–74 — **216**

KSU makes claim of ownership	220
The back story to KSU's claim	221
A trail of lawsuits	226

16. THE TWISTED NEMATIC LAWSUIT: **NOVEMBER 1974 TO JANUARY 1976**	**230**
The trial begins	233
Making money while the trial goes on	236
Depositions and back to court	236
The judge works to settle the case	240
Back at home, Susan Michelle is born	243
The case is settled, though Jim remained uncredited	243
Hoffmann-La Roche pays license fees … for a while	247
17. STARTING THE AMERICAN LIQUID XTAL CHEMICAL CORPORATION	**249**
Making money from LC's—anything legal would work	252
Jim has a heart attack	253
ALX thrives	254
Recycling of their toxic waste	257
Scores of new applications using LCs get underway	259
Funny, but dangerous mishaps	262
18. ALWAYS THINKING OF A NEW IDEA	**265**
SM does not replace TN-LCDs but finds other uses	269
Inventing the ENCAP and another patent race	273
19. MOVING TO SILICON VALLEY	**277**
The formation of a new company, Taliq	281
The Vari-lite vision panel	284

Formation of OSI	286
Jim lobbies for inventors to have longer patents	287
20. THE TEKTRONIX LAWSUIT: A SET OF PIRATES	**290**
The jury decides in favor of Jim	295
21. MR. LIQUID CRYSTAL	**297**
And more inventions	299
Lobbying to change patent law to favor inventors	304
Jim's philanthropic donations	306
A lifetime of achievement and many awards	307
His heart kept ticking	309
Writing his own book	310
Jim's last years	311
Forever Mr. Liquid Crystal	314
ENDNOTES	**316**
RESOURCES	**323**
ACKNOWLEDGMENTS	**329**
ABOUT THE AUTHORS	**332**

INTRODUCTION

JAMES LEE FERGASON was the first researcher to explore the practical applications of liquid crystals (LCs). He invented the twisted nematic liquid crystal display (TN-LCD), ubiquitously used today in the screens of smart phones, calculators, iPods, flat-screen TVs, digital watches, laptop and tablet computers, medical equipment and many other devices. If you press your finger against any LCD panel, you will see a change in color due to a change in the alignment of the liquid crystal molecules that are a component of the display. Electricity changes the orientation of the liquid crystals to conduct light, allowing you to see the image.

This is the untold story of how Jim Fergason created interest in liquid crystals and ultimately changed the way we do many things, from telling time, to watching stories, to communicating information to each other.

Born on a small Missouri farm during the Great Depression, he became a pioneer of liquid crystal optics. Among the top inventors in the history of the U.S., Jim held 150 U.S. patents and 500 foreign patents in the field. The patent examiners at the U.S. Patent and Trademark Office (USPTO) called him "Mr. Liquid Crystal" and he was treated like a rock star by fellow inventors, though his name is unknown to the general public.

His concept for the TN-LCD was only the starting point of the applications of this technology. As he often said, "An idea is not the same as an invention." Making a viable commercial liquid crystal display required developing a system of interacting elements, each with particular optical, electrical, and mechanical properties. Other researchers concerned that a display utilizing optical filters called polarizers would not be bright enough had rejected their use, but Jim explored polarizers, sought improved types and invented a unique, non-depolarizing reflector located behind the rear polarizer that is twice as efficient as the conventional reflecting surface. Jim invented the spacer seal, a thin polymer film that holds the two pieces of glass in a display together and acts as a barrier preventing the liquid crystal from seeping out and moisture from seeping in. He also developed the most extraordinary component, the surface alignment coating, a very thin, oriented polymer film that guides billions of molecules in the liquid crystal layer to align into a one-quarter twist.

Additional inventions he made using liquid crystal technology include those that dramatically increased the switching speed of liquid crystals, which opened doors to new applications like 3D movies and rapid motion images. He also invented the nematic curvilinear aligned phase device, which consists of micro-droplets of nematic liquid crystals with unique optical and electrical properties that are encapsulated in a flexible polymer film to make architectural windows that look like opaque frosted-glass shower doors but can then be electronically switched on to suddenly turn clear.

Jim was one of the last independent American inventors to follow in the tradition of the greatest independent inventors such as Nikola Tesla, Edwin Land, George Washington Carver and Edwin Howard Armstrong, all of whom thrived at the turn of the 20th century. Like them, he worked mostly on his own in small laboratories he funded himself over his lifetime, rather than in industrial corporations.

For his achievements, Jim was inducted into the National Inventors Hall of Fame. He received the Lemelson-MIT Prize, the Jun-Ichi Nishizawa Medal and many other prestigious awards, but he often donated his prize money to support independent inventors and to fund science scholarships for college students. He was always generous in sharing his knowledge with others, and educated the people who worked with him by constantly discussing various liquid crystal theories and ideas for devices, and using his broad technical knowledge and good judgment to guide their work. Like two Missourians he admired, Mark Twain and George Washington Carver, he remained humble and approachable even after achieving fame within the field he loved. When the University of Missouri awarded him an honorary PhD in 2001, the new "Dr." Fergason told a reporter, "Don't call me Doctor. The only doctorate I ever earned was in the school of hard knocks."

HISTORY OF THIS BOOK

This is the story of how an independent inventor persisted against scientific competition, numerous international patent disputes, corporate power and near financial disaster to invent all of the major technology enabling the modern liquid crystal display industry.

This book is based on the original "autobiography" that Jim Fergason was writing before his death in 2008, which was co-authored with his close friend, Arthur Berman. In 2010, Jim's daughter Terri Fergason Neal decided to complete that manuscript and publish it, largely for the Fergason family but also for the many scientists, inventors, and other interested persons who would enjoy knowing more about the life of a leading American inventor. Terri worked closely with editor Marian Pierce over the next five years to expand the manuscript and finalize it.

Together they interviewed and emailed extensively with Fergason family members, as well as many of Jim's colleagues, friends, and supporters, especially Tom Harsch who worked closely with Jim for more than 40 years and whose familiarity with the events and history added extensive detail to support the original manuscript. Additional inputs from Dora Fergason, Jim's wife, and Terri's siblings (brothers Jeff and John, who were long-time leads in the family ventures, and sister Susan) helped complete the materials.

One of the key elements of this book is to set the story straight about how Jim Fergason was, without any doubt, the true inventor of the TN-LCD. As this book will show, he was the first to utilize and exploit the science of liquid crystals, to manufacture them, and to apply them for inventive, practical purposes. He was credited by the U.S. Patent Office with the first patent on them. He is Mr. Liquid Crystal.

1. FROM A SIMPLE BEGINNING

> *"Life on the farm was not all that different from camping, but without the luxuries."*
> —Jim Fergason, on his rural upbringing

JIM WAS BORN on January 12, 1934 at his parent's hilltop farm off of Coal Holler Road, near the small town of Wakenda in northwestern Missouri. He was a breech birth, delivered safely by a doctor who traveled to the farmhouse in -8°F temperatures. At age 41, it was a surprise late pregnancy for Jim's mother, Sarah Margaret Cary. Her next oldest child, Lewis ("Bub" for "Little Bubber"), was then ten years old, her daughter, Mary Margaret, was 18 and her oldest child, Edmund, was 20.

MATERNAL ANCESTORS

Jim's mother had taught in a one-room schoolhouse before her marriage. She was good at math, played piano, wrote poetry and composed plays for her students to perform. Her brown eyes shone with intelligence out of a round face with strong features and she always wore her long brown hair pinned up. She was tall for a woman and caught the eye of handsome, blue-eyed Joshua Fergason at church. The couple courted by horse and buggy. The horse was an asset to the courtship because he knew his way home and Joshua could let go of the reins when Sarah Margaret was in the buggy and hold hands with her instead. The couple had married in 1912 and moved to Wakenda after losing all their stocks in the Great Depression.

Sarah Margaret's ancestors had emigrated from Ireland to Virginia in about 1640. Her great-great uncle, Nathaniel Cary, had been one of the first white settlers in Wakenda in 1818. His brother, Sarah Margaret's great-grandfather, Hardin Cary, traveled from Tennessee to Missouri ten years later and settled near his brother on the banks of the Wakenda River. Hardin's son Daniel Hardin Cary, born while his parents were en route to Missouri, prospected for gold during the 1849 California gold rush but didn't strike it rich and returned to the family farm on the Missouri prairie. Daniel married, acquired more farmland, built a grand brick house in 1860 and was later elected a county judge.[2] During the Civil War, the Union Army staff held a party at this house that was attended by General William T. Sherman. The general got drunk and kicked a hole in the plaster with his spurs. (Jim recalls that, as a child, he was looking through the wreckage of the Cary family house which had been destroyed by a tornado and found the hole, which had been patched with plaster, in one of the thick brick walls still left standing.)

Daniel's son, e.g., Jim's maternal grandfather, William T. Cary, was born during the Civil War, when the state was a tangle of Union Red Legs and pro-Confederate rebel guerrillas like William "Bloody Bill" Anderson's raiders. William's mother occasionally had to hide him in the cistern to keep him safe. The homestead was known locally as the Cary Stock Farm. In an era when few attended college, William was a bright young man and obtained a college diploma at the age of 15. He became a Methodist minister known as Parson Bill with a circuit ministry. Grandpa Cary and his wife, Minnie Bell Winters Cary, were both alive when Jim was small and he saw them often.

Jim's father and mother's wedding portraits:
Joshua and Sarah Margaret Fergason, 1912

PATERNAL ANCESTORS

On his father's side, Jim's ancestors had emigrated from Edinburgh, Scotland to Virginia in the late 1850s. Jim's paternal grandfather, Edmund Ferguson, had a thriving farm in Indiana in the late 19th century. Edmund was illiterate and liked to play cards and drink whiskey. According to family lore, he lost his farm in a card game, moved to Missouri seeking new opportunities and then sent for his wife, Lucy Ann. At first she refused to come, but her family eventually persuaded her to. When she left Indiana, she said she would never return.

Lucy Ann wouldn't let Edmund gamble or drink alcohol in Missouri. Edmund became a successful farmer and taught

Jim Fergason at 3 years old, 1937

himself to read by studying the *Wall Street Journal*. Edmund's oldest son, Jim's Uncle Charlie, accidentally changed the family name while registering for school by replacing the "u" in Ferguson with an "a" to make it Fergason.

The farm where Jim spent his childhood was 120 acres, of which only about 40 were tillable. As a small child, Jim explored the land carrying his black stuffed dog on his forays. He had his mother's brown eyes, ears that stuck out and a sunny disposition. His family called him Jimmy.

LIFE ON THE PRAIRIE

The farmhouse had no electricity or indoor plumbing. Water came from a pump about 100 yards from the house and was naturally saturated with calcium carbonate. Since such hard water made for poor cleaning and laundry, Sarah Margaret collected rainwater in a cistern and used it for washing. Everyone bathed in a galvanized tub. With no indoor plumbing, fair weather or foul, it was off to the outhouse whenever the need arose. However, the house was connected to the community by a party line telephone.

Winter nights were so cold that the glass of water Jim placed by his bed would be frozen by morning. Heat came from a stove in the main living area. At night Joshua loaded the stove with coal and closed the dampers to limit the air supply, which made the fire smolder all night long. When Joshua got up in the morning to feed the animals, he opened the damper and added dry corn cobs. The stove soon had a roaring fire. By the time Jim climbed out of bed, the stove was glowing red and spreading warmth.

The Fergasons hunted, raised, or gathered almost all of the food they ate, except for staples like flour. Joshua tried to raise cattle, sheep and hogs. Once a year the family butchered a hog and rendered the lard, which they used all year for cooking. For breakfast they ate salt pork, scrambled eggs and brains with biscuits or homemade bread. Sarah Margaret had chickens and sold the eggs for 6 cents a dozen. She canned fruits and vegetables, storing them in the root cellar. Much of the Fergason's sweetener came from honey and it was a matter of family pride that Grandpa Cary was a beekeeper. From an early age, Jim did his part to help fill the larder. His mother taught him how to forage for edible plants like lamb's quarters, which tasted like spinach; wild dock, a member of the buckwheat family; wild lettuce and dandelions. He also learned to identify poisonous plants like foxglove and night shade.

Like the other children in the area, Jim was expected to help with farm chores after school. The family took Sundays off to attend church. They were practicing Methodists and Sarah Margaret taught Sunday school. Their religious beliefs were a quiet but constant part of their lives.

The farm had no refrigeration and fresh meat was an important food source. Sarah Margaret taught Jim how to shoot a .22 rifle when he was just five years old. At the time, the rifle seemed enormous. The Fergasons hunted and ate deer, geese, quail, squirrels and wild ducks. Joshua was a good hunter but he didn't hunt for sport. He taught his children never to waste ammunition shooting game that wouldn't be eaten. The family also fished from the local stream.

Jim's parents were described as "incredibly kind" by everyone who knew them. Joshua was a disciplinarian, smart,

Jim and his parents, 1944

jovial and an ace checkers player. He was never successful at farming but with his full-time job as the postmaster of the Wakenda Post Office, the family had security during the hard times of the Depression. Postmaster jobs throughout the nation were obtained by political patronage. Joshua was a Democrat, appointed postmaster in about 1933 during the Democratic administration of Franklin D. Roosevelt. Undertaking the postmaster job required purchasing the assets of one's predecessor, including the mail boxes. Like many in the country, Joshua anticipated a defeat for Truman in the 1948 election and gave up his postmaster position to work as a mail carrier in Carrolton. Jim would ride along in the mail truck with his father sometimes and take a shot at any rabbit he saw.

JIM'S SIBLINGS

Jim was just two years old in 1936 when his oldest brother Edmund, who had been working the family farm, left to pursue a degree in electrical engineering at what was called Missouri University, or MU. Edmund's decision to attend college came at a difficult time for the family. Missouri had been hard hit by the drought of 1934, which covered almost 80 percent of the contiguous United States. Additionally, that year proved bad for crops, combining a dry season with no market. The family was broke. Joshua told Edmund it was a waste of money to attend college, but Edmund was determined. He left with only $20, which he had earned over the summer by doing extra jobs. The family didn't see him again until he came home for Christmas and placed $20 on the kitchen table, money he had saved by working after school and living frugally. This made Edmund a hero to the whole family.

Edmund participated in the Army Reserve Officer Training Corps (ROTC) during college, which earned him extra cash and a commission in 1939. In 1940 he went to Fort Hood,

Texas and was one of the first men in combat in World War II. Nicknamed the "Desert Fox," Edmund served as a major under General George Patton in the North African Campaign and afterwards in the Sicily campaign. His military career continued under General Mark Clark in the 1943 Italian campaign. He went with Clark to Austria in 1945, where Clark accepted the surrender of German troops. Edmund became the "mayor" of a small Austrian town. Jim looked up to Edmund because he had fought for his country.

Jim also admired his next oldest brother, Lewis, but Lewis wasn't always the best role model. When he was a child he put a firecracker in a bottle and the bottle exploded, shooting glass in his arm. He once took apart his mother's sewing machine, removed the Briggs and Stratton gasoline engine from the washing machine and used the engines from the two different machines to make a power jigsaw. His parents were exasperated, but Lewis had a way with words and could always talk himself out of trouble.

Lewis started college at Missouri University in 1941, majoring in chemistry. Jim was just seven years old then, attending the Rosebud School, a one-room schoolhouse for children in grades one through eight. At school Jim learned about famous inventors like Thomas Edison, Alexander Graham Bell and George Westinghouse, but it was his brother Lewis' experimentation and interest in how things worked that influenced Jim to become a scientist and inventor. In 1989, Jim told a reporter that "Lewis invented his own formula for gunpowder," "designed small loudspeakers" and was "always using his devices to play practical jokes."[3]

JIM'S YOUTH AND INTEREST IN SCIENCE

Growing up, Jim didn't have other children his age to play with and filled this void in companionship with dogs, cats, turkeys, ducks, rabbits and a lamb. He also made pets out of orphaned wild animals, including foxes and squirrels, and served as the family veterinarian, caring for generations of injured chicks.

Jim once tried to make a pet out of a goat. As an adult, he could still recall his disappointment over how this turned out. He earned $10 working all summer gleaning corn from around the grain elevator and used $4 of this to buy a billy goat from a farmer, thinking it would make a wonderful pet. Jim put the goat in the same fenced-in area of the yard as his mother's chickens. The chickens weren't able to get out but the goat bounded over the fence and ate all the flowers in the garden, dragged the clean laundry off the clothesline and stomped on it. Jim's father ordered him to get rid of the goat, so Jim led it by rope back to its original owner. After sharp haggling, the farmer agreed to take back the goat if Jim gave him $5. In a lifetime of business dealings, Jim regarded this as his first experience at having to start over again.

Although Jim learned to ride a bicycle, he was uncoordinated and couldn't ride through a farm gate without hitting both sides. However, he had Herculean strength and excelled at watermelon chucking and tossing his little nieces Diane and Carol Jean over his head. Carol Jean remembered Jim telling her not to go into his room because there was a "Hippadankus" in there, a horrible monster that eats little children. One time, Jim "fought" the Hippadankus in his room.

He appeared after the fight, red-faced, sweaty and carrying a large fake sword. Over time Carol Jean realized the Hippadankus was a make-believe creature her uncle used to keep nosey little girls out of his room.

Jim and his nieces, 1947

Jim's rural upbringing shaped him. The classic children's book *Two Little Savages* by Ernest Thompson Seton was Jim's favorite when he was a child and the life of the book's hero Yan, with his insatiable curiosity about science, animals and the natural world, mirrored Jim's own. Yan, a city boy, goes to stay with a friend in the country. The two boys learn how to observe nature without interfering with it. They learn to live in the wild, relying on American Indian lore and the ability to identify medicinal and edible plants.

For recreation, the Fergasons played horseshoes and croquet and fished for small brook trout in a stream on the land. Benson's quarry, which had once produced white limestone for use in the construction of local buildings, was also on the farm. By the time Jim was born the quarry had become a deep swimming hole filled with "sweet" water saturated with calcium carbonate from the quarry walls. The Fergasons swam in the quarry but preferred the pure waters of the stream, fed by a natural spring. Long after the family moved, the spring was dammed, resulting in the formation of a lake that now covers the entire property.

Jim said of his experience growing up in rural Missouri in the 1930s and 40s:

> [S]uch a lifestyle provides clear lessons to those that live it. Nature doesn't care what people think. The laws by which nature runs the world are not decided by taking a vote. You play by nature's rules. Magic may be fun but does not lead to food on the table. Nature requires people to learn and understand their environment the way it really is. Only by doing so can people survive and enjoy life. Prior experience and logic are the keys to survival.
>
> Death was more personal. Smallpox had been eradicated, but people still died of whooping cough, measles, rheumatic fever, polio and other diseases that are now prevented by vaccinations or treated by antibiotics. Infections were common. A neighbor whose hand was punctured by a small splinter from a fence died of tetanus. At that time, people knew the alternative to progress—up close and personal.

In 1945, when Jim was 11, the family moved eight miles up the road to property they had bought in Carrollton, Missouri, a town of 4,000. Sarah Margaret grew a huge garden, wrote poetry and had an active social life. She joined the Daughters of the American Revolution and worked as a substitute schoolteacher. The students liked her because she was a good storyteller.

Jim had a favorite pet at his home in Carrollton, a feisty fox terrier named Rascal who had lost an eye in a fight with a cat. Rascal hung out at the neighborhood bar because he liked beer and some of the customers would give it to him. One day he was ambling home drunk from the bar and a car struck him on his blind side, killing him instantly. The bar owner felt terrible. Nobody had extra money in those days, but the bar owner was fond of Rascal and sacrificed a new sheet to wrap the dog's body in.

After the move to Carrollton, Jim became very curious about science, and by high school, he had read all of Lewis' chemistry books on his own. He also read a book on rocket propulsion that described a chemical reaction that produced a high specific impulse and wanted to try duplicating this chemical reaction. His "experiment" filled the school gym with heavy, white smoke, so thick it was impossible to even find the windows. As punishment the athletic coach failed Jim in physical education class that term and Jim had to take it over for an entire year, which was hard for him since he didn't like sports. The middle finger of his right hand was permanently bent from one of the few times he had played football as a child.

Jim spent the summer of 1948 working for the Pfister Hybrid Seed Corn Company hoeing weeds from rows of corn that averaged one-mile long, a process called "laying by."

He was good at the job and the company promoted him to the crew that cut volunteer corn (residual corn from previous season) to prevent cross pollination. Later, they put him in charge of stripping the tassels off the corn so it didn't self-pollinate. By the end of a summer of this grueling work, Jim knew that he didn't want to become a farmer. He'd go to college if he could and try to turn his interest in science into a career.

Throughout high school Jim worked at the Smith Rexall drugstore to earn spending money and to help out his family. He washed windows, stocked shelves and delivered drugs to doctors around town. His least favorite chore was cleaning the bathroom.

MEETING HIS FUTURE WIFE, DORA

In the summer of 1951, before Jim's senior year of high school, a friend named Paul Minor took Jim after work to a local teen hangout, the soda fountain at Roseberry's Drugstore. Paul introduced Jim to a girl he was interested in dating, Martha Ann Benson. She had brought along a classmate named Lorita Sue Griffin, plus a girl Jim didn't know, Dora Delaine Barlish. Dora, a pretty, vivacious 14-year-old, sat across from Jim at the table. She recalled thinking how tall he was and remembered that he kept his big brown eyes on her throughout the evening.

A week later, Martha and Paul wanted to go to a Sunday matinee at the Uptown Theatre, which showed the best movies in town. Teenagers in Missouri then weren't allowed to go on dates alone and had to either double-date or bring along a chaperone. Martha's mother called Dora's mother to ask if

Dora could come along, and Paul brought Jim. Paul and Martha had an ulterior motive for matchmaking. Paul didn't have a car, but Jim was allowed to use his father's green Chevy. If Jim dated Dora, Paul and Martha could tag along in the back seat.

This plan worked to Martha and Paul's advantage but not to Lorita Sue's. She had expected Jim to invite her to a school dance, but he had fallen for Dora the first time he laid eyes on her and invited her instead. Dora recalled telling her mother that a young man named Jimmy had invited her to a school dance. Dora's father Ezra said teasingly, "I'm happy you're not going to be an old maid," while her mother Fern was pleased because she thought her daughter was referring to another Jimmy in town, someone from a family that had more farmland.

Dora and Jim made an attractive couple. Jim was six feet one to Dora's five feet two, and had wholesome good looks, a warm smile and a hearty laugh. Dora was a ladylike blonde who could sew absolutely anything and was an excellent cook. She found Jim's friendliness and determination to attend college appealing. Despite her domestic skills, she knew, even at 14, that she didn't want to be a farmer's wife. She didn't particularly like animals or enjoy getting dirty.

Their backgrounds were similar in other ways, although Dora had grown up on a more prosperous farm than Jim. Her father, Ezra, had been a sharecropper but after acquiring a 600-acre farm outside of Bogard, Missouri, he became a gentleman farmer, raising chickens, hogs and cattle and growing corn, alfalfa and hay. Every farm needed a riding horse to herd the cattle from pasture to pasture and Ezra rode a spirited, five-gaited steed named Lady Charlemagne that Dora De wouldn't go near. The horse, nicknamed Sugar, had been raised on the J.C. Penney horse farm in Hamilton, Missouri. Dora's mother, Fern Barlish, was a practical person with abundant common sense. She gardened, raised chickens, sewed quilts, embroidered and passed on a love of flowers to her daughter.

High school photos—Jim Fergason (1952) and Dora Barlish (1955)

COLLEGE YEARS

Jim considered studying chemistry at college, but his high school physics and math teacher, Hazel Shelton, encouraged him to major in physics and minor in math. When he graduated from high school in 1952 as part of a graduating class of 54 students, he was awarded an O.M. Stewart Scholarship in physics at Missouri University in Columbia. The scholarship and in-state tuition helped the Fergasons keep down the cost of Jim's college education. Still, Jim's father wasn't sure they'd be able to afford the entire four years.

Jim tested out of college-level organic chemistry and had confidence in his academic abilities, but felt like a country bumpkin among the city sophisticates from St. Louis and Kansas City. He had an advantage over other students in that his two older brothers had both graduated from Missouri University. They helped Jim adapt both socially and academically, and he enjoyed his college years.

Physics was a tough major and math a demanding minor. In addition to hours of study, Jim enrolled in the ROTC. At MU this meant the field artillery and, in his last two years, a little extra money. Jim also worked in the physics department as a lab instructor, grading an endless stream of student papers. In one of his favorite non-science classes, on the history of filmmaking, he developed an artistic appreciation that later helped him understand what people liked best about liquid crystal displays.

Carrollton was just 90 miles from Columbia, but Jim was so busy with his studies and jobs that he could only make it home to see Dora and his family every six weeks. He got rides from other students or hitchhiked, always toting his laundry. He joked that Dora's father liked him because he wasn't around enough to be a nuisance and that his romance with Dora lasted for the same reason. Dora was occasionally able to come to campus for school dances, and the couple's relationship had further opportunity to grow during the summers, which Jim spent back in Carrollton.

Dora graduated from high school when she was 17 and moved with her friend Martha Ann to Kansas City to work posting payments for Kansas City Life Insurance. She still saw Jim and during his last year of college, he proposed. Before Christmas of that year, he bought her a ring. His mother thought he was too young to get married. Dora's mother wasn't thrilled either because her daughter was only 18, but she said that Jim appeared to be smart and so she gave her approval.

FIRST PROFESSIONAL JOB

Jim had thought about going to graduate school, but decided to get a job to support Dora instead. His physics professor wrote a letter of recommendation attesting to Jim's excellence as an experimentalist and he used this letter to look for a job. He got offers from two premier companies with industrial research labs, Bell Labs and the Westinghouse Electric Corporation. He chose Westinghouse for many reasons, but the deciding factor was the salary offer of $400 a month, $10 a month more than Bell Labs.

Dora worked at the life insurance company until May 31, 1956. Jim graduated on June 6, 1956. Three days later, he and Dora drove in the first car Jim had ever bought, a light blue Chevy with a straight shift, on a gravel country road to Jim's parent's house in Carrolton. There, the plan was to get the car cleaned and polished so that after the wedding ceremony the next day, the bride and groom could drive off into the sunset. Along the way, another car sped over a hill and crashed into the Chevy. The impact caused the back seat to fly up and hit the roof. Remarkably, the couple wasn't injured, even though cars in those days weren't equipped with seatbelts. The car was wrecked. The couple immediately bought a second car, a turquoise and white Chevy with a rooster tail on the back. They married the next day and drove to Eureka Springs, Arkansas for their honeymoon.

Jim and Dora's wedding photo, 1956

Soon after, Jim was commissioned as a second lieutenant in the army, but didn't immediately have to report to duty because he had started at Westinghouse and the military deferred the start date of duty for scientists and others who were doing important work.

2. A WESTINGHOUSE ENGINEER

> *"To have a career as an inventor,
> an open mind is an absolute must."*
> —Jim Fergason

AFTER HONEYMOONING IN Eureka Springs, Arkansas, Jim and Dora drove to Pittsburgh, Pennsylvania, but Jim's hometown newspaper erroneously reported the newlyweds went to Pittsburg, Kansas. The couple had planned to arrive at 3 PM so they could relax and start learning their way around, but didn't realize that Ohio was on standard time and Pennsylvania on daylight saving time. They hit downtown at the height of rush hour. The parkways, which today give easy access to the city, had not yet been built. Coming from the west in rush hour traffic was like crossing a bridge to hell. In 1956, although air pollution from the steel mills had decreased due to smoke control laws, coal soot blackened every building. It was clear how the Smoky City had earned its nickname.

The Fergasons drove through the Hill District, crossed the University of Pittsburgh and crawled through the 4,225-foot-long Squirrel Hill Tunnel, a claustrophobic experience. Appalled by the traffic and filth, Dora wanted to retreat to Missouri. Instead of stopping, the couple drove on and spent their first night in Monroeville. The next day they went to Erie and stayed at a motel on the lake, enjoying strolls on the beach. A week later, they reluctantly returned to Pittsburgh.

Westinghouse's new employees attended a six-week orientation program, where they learned about the different divisions within the company. Representatives from the jet engine plant, the nuclear energy division, the research lab, which came up with new product ideas[4] and the central research lab spoke about the work being done at their facilities. After the representative from the central research lab described all the marvelous things underway in the laboratory, Jim knew he wanted to work there. However, most of the staff had PhDs and Westinghouse didn't assign new hires to that division.

Jim requested assignment to Westinghouse's Elmira electronic tube division instead. For two months, he worked in an advanced design program related to electron image tubes. The manager thought the central research laboratory might be interested in hiring Jim and sent him over to be interviewed by John Coltman,

John the inventor of the scintillation counter and x-ray image amplifier was the head of electronic imaging and display department at Westinghouse. He fired questions at Jim on electron physics, just what Jim had been doing for the last two months. "I couldn't have planned it better," Jim said. "It was as though I had crammed for his questions. I think he was prepared to continue asking me questions until he found one I couldn't answer. John also called in Max Garbuny and told him that I was the best person he had ever interviewed from the student program. Max immediately made me an offer."

The new college graduate became one of only five professional employees out of 205 at the central research lab with just a BSc. degree. In a 2004 interview, Coltman said that Westinghouse recruited most new employees from PhD programs, although a few new employees had bachelor's degrees and continued to study at night for advanced degrees. Jim was an exception to this, but Westinghouse never pressured him to attend graduate school. Jim always felt he learned more science and technology in his job than he would have in any graduate program.

By 1956, Westinghouse was a large, engineering-based company and one of the principle manufacturers worldwide of electrical machinery, employing about 120,000 people. As a remnant from the early years of the company, all professionals, including new hires, were designated engineers. Westinghouse entered the military electronics business during World War II. Government-funded scientific and technological research during the war led to advancements in plastics, radar, x-rays, bombsights and atomic energy. After the war, the company became a leading producer of nuclear generating equipment.

Although General Electric was at the forefront of the appliance business, there was also high consumer demand for Westinghouse's products. With their sponsorship of the highly respected TV drama series, Westinghouse Studio One, the company had a cultural impact on 1950s America. Americans had a sense of wonder about scientific discoveries in that era and science-oriented TV shows like the *Bell Laboratory Science Series*, the *John Hopkins Science Review* and *Mr. Wizard* were popular.

The company had been founded in 1886 as the Westinghouse Electric Company by George Westinghouse, an inventor and entrepreneur who received his first patent, for a rotary steam engine, when he was just 19 years old with support from another great inventor, Nikola Tesla. Tesla, a genius who loved electricity and was fascinated by lightening, was as famous during his lifetime as Thomas Edison and Guglielmo Marconi. He is best known for inventing a polyphase system of alternating current that is still used today. He changed life for us all, but died poor. This was an ironic beginning to Jim's career as an inventor.

THE INFLUENCE OF MAX GARBUNY

Max Garbuny, Jim's supervisor, was a brilliant physicist who had fled Nazi Germany. When Hitler came to power in 1933 and expelled Jewish students and professors from universities, Max was in the doctoral program in physics at the Technical Institute of Berlin. His mother was Jewish and his father Russian, but the Nazis considered him a Russian and allowed him to stay in school. After completing his doctorate in 1938, Max and his brother obtained U.S. visas. Just three months before the Nazis invalidated Jewish passports, the brothers left Berlin by train for Amsterdam. The Schutzstaffel (SS), Hitler's elite guard, walked them off the train at a town on the German-Dutch border. They expected to be "executed on the spot" after an SS officer found a photo of a bullet in flight in their suitcase, but to their relief, the officer said that anyone could find such pictures "in every textbook on applied physics" and let the brothers go. They were allowed to enter the Netherlands and from there traveled to England, where they boarded a ship to the U.S.[5]

Max became a post doc at the Institute for Advanced Study in Princeton. Founded in 1931, the institute harbored scientists and mathematicians who had been persecuted by the Nazis. Albert Einstein was appointed a professor there in 1933. Max loved telling the story of how he had gone to his first seminar at Princeton and found himself seated next to Einstein, although, to Max's eternal regret, they didn't discuss physics.

An avid competitive chess player, Max joined Westinghouse in 1944, patented 33 inventions during his long career there and authored *Optical Physics*. Westinghouse published a series of books on science and Max co-wrote two, *The Science of Science, Methods of Interpreting Physical Phenomena* and *Seven States of Matter*. Physics was his great love and he was constantly preoccupied with thoughts of it. This made him a menace in the laboratory. He once threw a lit cigarette into a vacuum flask of liquid oxygen—with explosive results. Luckily, the flask was made of stainless steel. A lab technician had to follow the absent-minded physicist around to make sure he didn't blow up the labs.

Max admired Jim's insatiable curiosity about how the world worked and considered the 22 year-old to be more of a peer than a subordinate. He assigned Jim to work with Tom Vogl, a solid state physicist and the principle investigator for the Fort Belvoir-sponsored research project on a photocathode. This cesium-bismuth photocathode was to be used in light-sensing applications such as vacuum phototubes and in a contact image converter, a device used by astronomers to observe objects in space with a faint light emission. William E. Spicer, who had been a graduate student in physics at Missouri University while Jim was an undergraduate, was researching the physics of cesium-bismuth photocathodes at RCA Research Laboratories and Jim knew about Spicer's work.

Jim worked on this project for several months, enjoying it because senior management left him alone. The outcome was that he filed his first patent application, for developing a means of sealing an infrared window. Westinghouse had a tradition of encouraging invention dating back to the days of George Westinghouse and gave cash bonuses to inventors as an incentive. Jim said of this, "I was only making $400 a month and you got 50 bucks for writing a patent disclosure and another 50 bucks when you filed it, so you could make a quick 100 bucks for doing a little inventing on your own. So I took good advantage of that and also learned about inventing and how to get useful ideas. That's where I got my start in the inventing parade."

Jim's pending ROTC obligation came due in May, 1957, and he travelled to Fort Bliss in El Paso, Texas with Dora, who was now pregnant. The 12–week course he took on surface-to-air missile technology helped him learn more about handling dangerous chemicals and filled gaps in his knowledge of microwave technology. After completing the course, he taught a platoon of 70 recruits how to fire a 40mm cannon and a .50 caliber machine gun.

In an "infiltration" course, held on a rocky field dotted with holes loaded with blocks of TNT, Jim got an unforgettable training experience. The exploding TNT left a bad taste in Jim's mouth. To make it more "exciting," machine guns were fired just above (or so it seemed) the butts of the men. The velocity of machine gun bullets is greater than the speed

2nd Lieutenant James L. Fergason of the 79th Infantry Division, 1957

of sound. The only way the men could tell a bullet was above them was by a loud crack, caused by an actual mini-sonic boom. They were not able to judge from the sound how high above them the round was, so they hugged the rocky ground. Jim cushioned his knees by taping sanitary pads to them.

FIRST BORN DAUGHTER...AND FIRST LIQUID CRYSTAL EXPERIMENTS

He had been in the military four months when Dora gave birth on September 1, 1957, to the couple's first child, a daughter they named Teresa Lee and nicknamed Terri. Terri's arrival cost her parents a bargain $12 because her father's military benefits covered the birth.

The Korean War was winding down and Jim was discharged early from the ROTC, although he still had to serve eight years in the reserves. The new parents returned to Pittsburgh in December of 1957 and stayed in a furnished apartment on Hill Avenue in Wilkinsburg, but upon discovering mice in the apartment, they moved to a duplex on Churchill Road.

Westinghouse had canceled the cesium bismuth photocathode program Jim had worked on before entering the Army. It was against company policy to terminate returning servicemen and so the company had to decide what to do with their bright young scientist. Max Garbuny asked Jim to explore new ways to perform thermal imaging, and gave him a list of possibilities.

"Although I didn't have a specific budget, I had learned to scrounge the laboratory for unused equipment," Jim recalled. "Since I was knowledgeable about crystal optics for optical activity, I was able to eliminate most of the possible ways for detecting temperature Max had listed, and settled in to study optical activity."

Chemical compounds that have the ability to "rotate" polarized light are called optically active compounds. Optical rotation describes the property by which asymmetric molecules rotate the plane of polarized light. Jim soon concluded that it would be impossible to make a useful temperature measurement and visualization device based solely on optical activity because it was so sensitive to material thickness that controlling random variations would be impossible.

He searched the company library and found a book called *Optical Rotatory Power* by T. Martin Lowry, an English chemist. The book had a chapter on optically active crystals. The chapter described how a scientist named Freidrich Reinitzer conducted original research on 'crystalline liquids' and observed optical activity in them.

This was the first time Jim had ever heard of "crystalline liquids," also called "liquid crystals." Liquid crystals were not mentioned in undergraduate or even graduate studies of chemistry or physics in the late 1950s, and there were few recent publications about them.

The idea seemed a contradiction. "In almost every way a crystal would seem to be the very opposite of a liquid," Jim later wrote. "The molecules in a liquid, although not as randomly distributed as those in a gas, are not arranged in any order that extends over a distance greater than a few molecules across. The molecules in a crystal, on the other hand, are fixed in a regular, three-dimensional array."

Fascinated, Jim read that liquid crystals had high optical activity. The "very large rotatory power" or optical activity was first observed in certain derivatives of cholesterol when they were in a liquid crystalline (mesomorphic) state.

The optical activity of liquid crystals was very sensitive to temperature. Max had assigned Jim to find new ways of thermal imaging, and based on what he read, Jim thought liquid crystals might be a promising material for thermal imaging. There were no commercial research labs anywhere in the world working with liquid crystals. That was about to change.

3. A UNIVERSE OF DISCOVERY

> *"When I returned from the service to my old job at Westinghouse in November 1957, I found that my project had been terminated even though it was against Westinghouse policy to terminate returning servicemen. That's when I became a successful inventor, because I discovered that there was a state of matter called liquid crystals. It opened a new universe of discovery to me."*
>
> —Jim Fergason

JIM JOTTED DOWN the citation Lowry had given for Reinitzer's original article in German and returned to the library to see what else he could find out about Reinitzer and liquid crystals. He learned that Friedrich Reinitzer was a botanist who had discovered liquid crystals in 1888 when he was working at the German Institute of Prague. He was trying to determine the melting point of a cholesterol-based substance in carrots, cholesteryl benzoate, and was puzzled when the substance appeared to have two melting points. At 145.5°C, the solid crystal melted into a cloudy or milky liquid. At 178.5°C, the cloudiness suddenly disappeared, leaving a transparent liquid.

As the material cooled, violet and blue colors appeared. These colors quickly vanished and the sample appeared milky but still fluid. On further cooling the violet and blue colors reappeared, but the sample quickly solidified, "forming a white crystalline mass." Reinitzer thought this behavior might indicate impurities in the material, but the behavior didn't change after he purified the material further.

Reinitzer wrote a letter to an expert in crystal optics, German physicist Otto Lehmann, for help understanding this substance. He enclosed samples of cholesteryl benzoate and another cholesterol derivative that exhibited the same behavior, cholesteryl acetate. These substances "exhibit striking and marvelous apparitions that I do hope will be of interest to you," he wrote to Lehmann.

Lehmann examined the cloudy liquid that formed when the solid crystal melted at 145.5°C under a microscope and became convinced that it had a unique kind of order, a crystalline molecular structure, whereas the transparent liquid at the higher temperature had the characteristically disordered state of all common liquids. Scientists had long believed that there were only three states of matter: solid, liquid and gaseous. Lehmann realized the cloudy liquid was a new state of matter and named it "liquid crystals" because it had the optical properties of a solid and the physical properties of a liquid. In his 1889 paper "On Flowing Crystals," he describes the viscosity of the cloudy liquid as "soft, syrupy, and gum-like."[6]

Lehmann extensively investigated liquid crystals for over 25 years and wrote four books on the subject. Lehmann observed that liquid crystals were sensitive to light, sound, heat, mechanical pressure and other variables. Cholesteric liquids changed color in response to temperature, but Lehmann never envisioned a practical use for this property. Other scientists studied liquid crystals in the late 19th and early 20th centuries, but didn't investigate any practical applications they might have.

Researching 75 years after Lehmann, Jim described the physical and optical properties of liquid crystals in an article he wrote for *Scientific American* in 1964: "Mechanically, these substances resemble liquids, with viscosities ranging from runny glue to 'solid' glass. Optically, they exhibit many of the properties of crystals; for example, a typical liquid-crystal substance scatters light in symmetrical patterns and reflects different colors depending on the angle from which it is viewed. It is sometimes helpful to think of the liquid-crystal phase of matter as consisting of one- or two-dimensional crystals."

Jim also noted in an article he wrote about liquid crystals in *Electro Technology* that, "No practical application was made of liquid crystals until the late 1930s, when John Dreyer discovered that liquid crystals would align on a surface when rubbed." Dreyer had experimented with liquid crystals formed from dichroic dyes, which are dyes that show two colors. Within a few years, he developed a method of making linear polarizers on a thin film. First, he repeatedly rubbed the film with a soft cloth. Then, he mixed methylene blue dichroic dye with a volatile substance, methyl alcohol. The methyl alcohol dissolved the liquid crystals without reacting with them, leaving a residue of dichroic dye molecules on the film. These molecules aligned in the rubbing direction. The thin films prepared in this manner were linear polarizers. Polarizers are still successfully made using Dreyer's method.

A polarizer is an optical filter. An old *Anchor Optics* catalogue explains how a polarizer works: "Light is made of electric and magnetic waves at right angles to each other, emanating in all possible planes relative to the source. Polarization at this point is considered random and has no preferred direction. If a polarizer is inserted, the planes in which the electromagnetic waves travel can be controlled such that only a single plane exists. This is clearly shown when two polarizers are used in conjunction with each other. If the transmission axes (polarized planes) are parallel, the light will be transmitted. If they are perpendicular, no light will pass."[7]

Put another way, light travels in a straight path, but vibrates in all directions. A linear polarizing filter causes the light to vibrate in one single plane by eliminating the other directions for the light to travel. It works like a venetian blind, allowing the light to pass through in only one direction.

Liquid crystals are elongated molecules that exhibit periodicity. Lehmann originally divided liquid crystals into two classes: crystalline liquids and liquid (or flowing) crystals, but Georges Friedel, a scientist who had investigated liquid crystals from 1907 to 1931, later reclassified them into three "phases" rather than classes: smectic, nematic and cholesteric.

- Smectic liquid crystal molecules are the most ordered. "Smectic" is the Greek word for soap, and soap bubbles are an example of a common smectic liquid crystal substance.

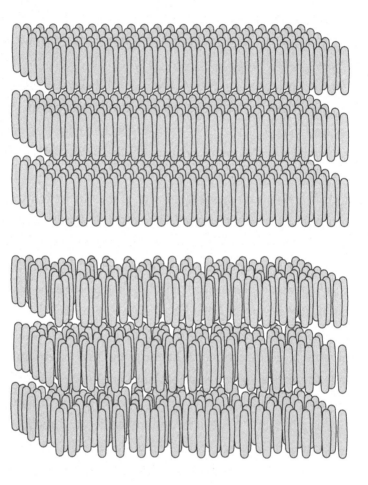

FIGURE 1. Smectic Liquid Crystals are either arranged in orderly rows or at random with the molecular axis perpendicular to the substrate in a series of layers. The layers are free to slide over one another giving the substance the mechanical properties of a fluid.

- The long axes of nematic molecules, like smectic liquid crystal molecules, are arranged parallel to one another and "stand up" perpendicular to the layer plane. However, the molecules are not ordered in layers. Nematic liquid crystals move freely, like a liquid, but are oriented in one direction. Friedel named them "nematic" for the Greek for "thread" because they contain microscopic thread-like structures. Jim described the molecules as behaving like a "long box of short toothpicks, which are all free to roll around and slide back and forth but which remain parallel to the long axis of the box."

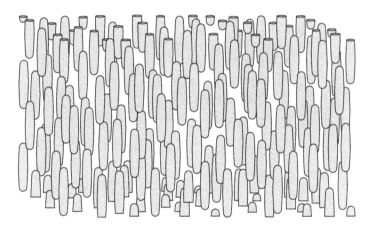

FIGURE 2. Nematic Liquid Crystals molecular axes are also arranged perpendicular but are not separated into layers.

- Friedel named the third class of liquid crystals cholesteric because their molecular structure is characteristic of numerous compounds that contain cholesterol, although cholesterol by itself doesn't have a liquid crystal phase. In the cholesteric phase, the liquid crystal molecules are twisted and will scatter light if the spacing between the molecules is a wavelength long.

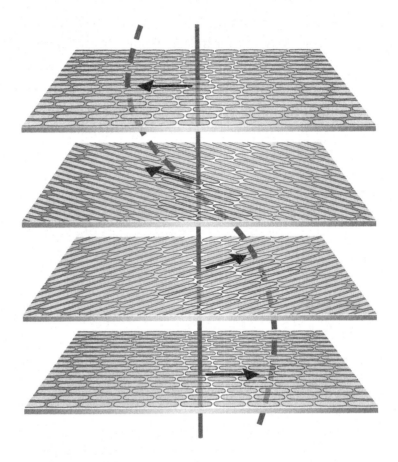

FIGURE 3. Cholesteric Liquid Crystals are arranged in layers similar to Smectic but the long axes of the molecule are parallel to the plane of the substrate. Because of the irregular shape of the molecule each thin layer is displaced slightly resulting in a helical path.

Jim summarized these in his *Electro Technology* article: "The nematic and cholesteric phases are liquids which have optical properties. The smectic phase is a true intermediate state between crystalline and liquid."

DIGGING DEEPER AND DISCOVERING POSSIBILITIES

In doing further research, Jim unearthed a paper by Reinitzer on cholesteryl derivatives, confirming Lehmann's initial observation that liquid crystals had a high degree of temperature sensitivity. He also found information on compounds of cholesteryl in a supplement to Elsevier's *Encyclopedia of Organic Chemistry Series III-V*, a book by R.P. Cook entitled *Cholesterol*, published by Academic Press in 1958 and in German dissertations from the Harvard library. Most of these dissertations were from the early 20th century. With his high school German, Jim tried to read as much as he could. There was a great deal of conflicting information but no indication how temperature sensitive liquid crystals were.

In *Colloid Chemistry* by J. Alexander, an article by Friedel indicated that a mixture of cholesteryl acetate and cholesteryl benzoate was stable at room temperature.

Jim decided to duplicate Reinitzer's original experiments and was able to obtain these chemicals from Aldrich Chemical Company in Milwaukee, Wisconsin, along with cholesteryl palmitate and cholesteryl chloride. The chemicals were made from cholesterol purified from wool grease and felt to Jim like "greasy kid's stuff."

He mixed the cholesteryl acetate and cholesteryl benzoate in a disposable aluminum weighing pan and heated them on a hot plate. He removed the pan and watched the clear liquid cool. As the chemicals passed through the liquid crystal phase, they flashed a rainbow of color, from vivid red to gold to green to blue to violet. The liquid crystal formed as a brightly colored iridescent—opalescent liquid. But liquid crystal molecules are not luminescent. Rather, the mixture was brightly colored by reflected light showing a spectrum of colors. It was love at first sight. "The colors might dazzle an impressionist's mind," Jim said.

Jim was immediately taken by the possibilities. Here was a substance which generated the beauty and appearance of opals and had clearly been neglected by the scientists who had first studied it. Jim decided to adopt it as his own. From that point on, he was like an explorer seeing a new country for the first time.

THE CHEMISTRY OF CREATING A ROOM TEMPERATURE LIQUID CRYSTAL

Jim observed that cholesteric liquid crystals exhibited intense light scattering and that the color of the scattered light varied rapidly with temperature. He wanted to investigate this temperature effect. In January of 1958, he launched a one-man project to develop a temperature measuring and visualization device based on cholesteric liquid crystals.

He faced an immediate obstacle. To start this project, he needed a cholesteric liquid crystal that was in its liquid crystal phase at room temperature in addition to being thermally and chemically stable. There were no materials with these properties.

Jim decided to make such materials himself. This was an ambitious goal for a chemist, let alone a physicist. Although Jim had shown a natural aptitude in his high school chemistry classes and tested out of college-level organic chemistry, designing and synthesizing molecules is an art and science both. Jim needed to learn more chemistry to begin.

At the library, he found the perfect textbook, the 1000-page *Practical Organic Chemistry including Qualitative Organic Analysis* by Arthur I. Vogel. One of the many glowing blurbs on the inside flap jacket stated that the book "would have my vote for the one organic manual to be stranded with in a desert island chemistry laboratory."

This praise was not exaggerated, Jim soon found. The textbook contained everything he needed to know to make organic compounds. He learned how to make precursors, which are chemicals that can be transformed into other chemicals or compounds through a chemical reaction, and how to make solvents, which are liquid compounds that can be used to dissolve gases, liquids or solids without reacting with them. He also learned how to make reagents, which are substances used in analysis and synthesis because of the specific reactions they cause. He even read the step-by-step instructions on performing glass blowing to make his own flasks for use in the chemistry lab, though he never attempted to make any. Chemists did not have any specialized apparatus or analytical equipment in those days, so it was easy for Jim to follow the instructions in the book. He only had to determine melting points and identify specific odors.

Jim felt confident that he was now ready to attempt to design molecules by using models and to synthesize new cholesteric compounds from simpler available chemicals. However, he had no funds to purchase the chemicals and equipment for the experiments. This was the high-tech equivalent of needing to make bricks without mud.

He again contacted Aldrich Chemical Company, this time to request free samples of cholesteryl acetate and cholesteryl benzoate. He begged colleagues so many times in other Westinghouse labs to lend him equipment until they either said yes or locked the doors when they saw him coming. Even a budget allotment to pay the bill he had accrued at the company library became an issue, but Jim persisted until he had everything he needed to start experimenting.

He still faced a major problem. All known cholesteric liquid crystals available at the time were solids at room temperature. This was because most chemicals were purified with a crystallization step at room temperature. In effect, any organic synthetic chemist would have had a mental block against materials that needed to be purified by other means, such as filtration. This type of situation occurs frequently in science, when moving forward requires an "inventive step," referring to the idea that scientists often need to proceed beyond a mental block. This expression was a favorite of Jim's, who often used it to describe his thought process. Jim remarked of this time:

> We were at a turning point. No one had previously considered the desirability of room temperature stable liquid crystals. This is obvious looking at the literature. There were thousands of compounds which were

synthesized without a room temperature material. There is a long list of compounds generated in the work by Jones and Gray emphasizing a property they called stability with high clearing points. In the International Critical Tables there were more than 20,000 compounds that had liquid crystal phases. There was no suggestion that stability could also be defined against the melting point of the crystals that formed or their ability to be chemically stable in the presence of oxygen or moisture. Note also that the scientists at RCA did not use room temperature liquid crystals in their first papers, but one of the most difficult chemical materials to work with, p-azoxyanisole. It was only after the Westinghouse liquid crystals became known beginning in 1963 that RCA hired Joseph Castellano in 1964. He advanced the development of room temperature LCs (liquid crystals). I felt like the one-eyed man in the kingdom of the blind.

The young scientist knew that two approaches could be taken to creating a room temperature liquid crystal. The first was the straightforward development of a pure material that is inherently a liquid crystal at room temperature. The second was based on the well-known effect in chemistry that when two or more chemicals are mixed together, the temperature at which the mixture crystallizes is lower than that of any of the constituent materials. By varying their proportions, an infinite number of mixtures can be prepared from even two materials. One of these mixtures will have the lowest crystallization point. This mixture is called a *eutectic*. Therefore, the second approach was to create a suitable eutectic.

Jim planned to pursue both approaches. "I began searching for a liquid crystal which would be liquid over a wide temperature range. The mixtures always tended to crystallize at some temperature and separate into their component parts. I started looking for a liquid crystal system which was temperature sensitive and would not separate. I also measured the optical properties in liquid crystals which had very little temperature sensitivity. In short, I took on the task of making a laboratory curiosity into a useful technology, which included finding and synthesizing the proper materials, finding ways to make them into thin films and discovering as many uses for them as I could."

Jim was trying to find a way to lower the temperature at which the chemical entered the liquid crystal phase. He reviewed the literature to learn everything possible about the shapes of molecules known to form cholesteric phases, then planned a series of experiments to determine what modifications in shape would result in a lowering of the liquid crystal transition point. His idea was to synthesize a series of liquid crystal molecules with the same general shape, but different electrical charge patterns or dipoles. An example of the changes in shape Jim had in mind was to create molecular isomers. These molecules have the same molecular formula and chemical bonds but have a different arrangement of atoms. An isomer is like a chemical anagram with the letters switched around.

Jim knew that he had a choice between active amyl alcohol and cholesterol as the means to synthesize liquid crystals. The most promising starting point was suggested by a 1938 paper on fatty acid esters of cholesterol. He pursued this approach and was able to create the first material that showed highly reproducible color play with strong temperature dependence, called cholesteryl nonanoate.

In previous chemical synthesis work, Jim learned two important characteristics: that cholesteric liquid crystals containing unsaturated fatty acids with a configuration called *cis* lowered transition temperatures, and that replacement of an ester chemical group in cholesteric molecules by a carbonate chemical group also lowered transition temperatures. Unsaturated bonds in chemicals are weak spots.

Making a carbonate required the use of a toxic chemical called *phosgene*, which was used as a chemical warfare agent in World War I. At a low concentration, the gas smells like new-mown hay. When used improperly, it can cause an explosion. There are few chemicals less user friendly and more dangerous than phosgene (plutonium being one). Jim had never worked with phosgene except on a poison gas infiltration course in the army, where a small amount of it was introduced so that the soldiers could learn to recognize the odor.

He knew in principle how to make carbonates using phosphate. Armed with this knowledge, he started to do his first reaction with phosgene. The phosgene had arrived in a steel container with two ports, one for removing liquid phosgene which boiled at 19°C (66°F) and one for removing phosgene vapor. What he didn't know was the extreme affinity and solubility of phosgene in benzene, his solvent. His plan was to add the liquid phosgene to the benzene. Instead of dissolving slowly into the benzene, the vapor immediately went into the solution with a liberation of energy, giving off heat and causing the solution to boil.

"The result was a very scary few minutes when I didn't know whether the safety features that I'd put in place were going to work," Jim said. "Fortunately they did. I immediately modified my apparatus to essentially inject the phosgene vapor into the benzene to form a solution and to use the solution as a reagent. I knew that phosgene smelled like new-mown hay, and my imagination was telling me that I was missing the odor! I will always remember how the phosgene that I took from the container as a liquid looked cold and deadly. I was thankful to have a safety plan. The system I developed to make carbonates gave very few side products, which made it easy to purify. For the next 14 years, I made cholesteryl oleyl carbonate with no incidents and very good results."

Jim considered cholesteryl oleyl carbonate to be among his greatest discoveries in his new role as a synthetic organic chemist. It turned out to be one of the most useful cholesteryl liquid crystal compounds because of its properties.

Jim needed the chemical nitroso-methyl urea to synthesize certain liquid crystal materials of interest. One of the most carcinogenic artificial materials ever created, nitroso-methyl urea is used to produce dinitromethane, an odorless gas that can kill a person who inhales it within several hours. The victim doesn't exhibit any symptoms after exposure. Under the wrong conditions, the gas can be explosive. The company purchased the chemical, but Jim decided it was too dangerous to undertake the synthesis.

His work usually had no impact his home life, although Dora occasionally found him distracted by work-related thoughts. But there was one exception. While investigating certain chemical derivatives, Jim synthesized materials called thio carbonates and aromatic nitrogen derivatives. The synthesis of thio carbonates starts with materials called mercaptans. Butyl mercaptan has the characteristic odor of a skunk. The amines had names such as cadaverine. Jim would go home with a deadened sense of smell and no perception that he stank. When Dora told him, he accused her of having a keen imagination. It wasn't until his mother visited and exclaimed, "Jimmy, you stink" that he finally understood.

JIM'S CHEMICAL SUCCESS

In the fall of 1958, Jim systematically synthesized a series of cholesteric compounds in order to increase his knowledge about them. The process of invention is sometimes romanticized, but often involves this kind of methodical approach. Jim eventually created 140 compounds, with Fred Davis helping with the final purification. As Louis Pasteur said, "Chance favors the prepared mind." Jim was lucky to have found the first room temperature material early in the program and to have been able to build a first class database after "only" 140 tries, but he achieved this goal through patient preparation, good planning, systematic and highly skilled experimentation and a keen analysis of the results.

"I made many mixtures and synthesized compounds that had not been synthesized before, and looked at them in ways they had not been looked at," Jim said of this experimentation. "Unfortunately, just as Friedel had said, the mixture would crystallize after a few hours, but I was able to make other mixtures that formed eutectic mixtures which remained liquid crystal at room temperature. This paved the way for many other practical applications, since it showed that it was feasible to make a very wide range of liquid crystals at room temperature that had utility. Liquid crystals would no longer be a laboratory curiosity!"

Forty-eight years later, when Jim received the Lemelson-MIT prize for invention, he said that while he didn't discover liquid crystals and wasn't the first to experiment with them, "I was the first guy who saw what they were really good for."

Despite the promise of Jim's work, it was an uphill battle to convince the management at Westinghouse of the possibilities of liquid crystals. Clifford F. Eve, a colleague of Jim's in the applied physics department, wrote that, "Jim's original entry into the liquid crystal field here was entirely the result of his own initiative, and I would say he received very little encouragement at all until he had gone far out on his own in demonstrating the importance and possibilities of this field of enquiry."[8]

Jim told a science reporter in 2002 that, "I had to get people to believe there was such a thing as a liquid crystal, even though it's colorful... and get them to believe that it was more than just a laboratory curiosity. And it took a lot of measurements and a lot of work. Then I had to show them not only that liquid crystal was important but that what you did with liquid crystal was important."

Over 50 years have passed since Jim first synthesized cholesteryl nonanoate and cholesteryl oleyl carbonate. They remain the most commercially successful cholesteric compounds ever invented. More of these materials have been produced than any other. Literally tons have been manufactured.

During this fruitful work period, Jim's father died. Jim remembered the impact it had. "In the midst of the euphoria of discovery, an unexpected event occurred. On November 4th, 1958 my father died instantly while working in his vegetable garden. He had hired a local man to prepare the fields with his horse and plow for spring planting. He had two potatoes clutched in his hand. His death created a hole in my life. He was not my friend or my pal but my dad who supported me and taught me the difference between 'can do,' 'make do' and 'will do.' My mother was facing a whole new life. They had shared life for half a century through good times and bad. In her bad times she turned to writing poetry for her own peace of mind, much of it on envelopes and scrap paper. After my father died she wrote this poem, which has inspired me many times when life seemed bleak."

Alone

Thoughts I now have are best left unsaid,
Lines I now write are best unread,
Too many tears are left unshed—
Still—
I saw a bug on my rug,
He had three legs gone,
He still crawled on.
The rug was rough,
The going tough,
He had strength enough
To creep on.

There was a bee on my windowsill,
He had just one wing, still

He had the will
To try to fly,
To strive to live,
And not to die—
I wonder why?

I feel akin to such as these,
Legless bugs and wingless bees,
For a great part of my life is gone,
Henceforth I must walk alone,
With no strong arm to lean upon.
But am I weaker than the bug
that I saw crawling on the rug?
Am I more spineless than the bee
that on the windowsill I did see?

Though my pathway will be rough,
Though the going will be tough,
God will give me strength enough
To walk alone.

—Sarah Margaret Fergason (1958)

4. THE WORLD'S 1ST LIQUID CRYSTAL-BASED DEVICE DISPLAY

"That is the essence of science: Ask an impertinent question, and you are on the way to a pertinent answer."
—Jacob Bronowski

ON AUGUST 1, 1959, a year after Joshua Fergason's death, Jim and Dora's second child, Jeffrey Keith, was born. Jim loved children and having a newborn at home helped assuage his grief over the loss of his father. He was somewhat unusual among scientists in that he worked hard at his career, but was able to balance its demands with family life. With two children, the young family needed more space and bought their first home, located above Rosedale, PA.

During the year between his father's death and Jeff's birth, Jim met a kindred soul, Wilhelm Stürmer, a chemical physicist and radiologist who was the research director at the medical x-ray division of the German company Siemens.

"Stürmer had an interest in the aesthetics of crystals melting and crystallizing in polarized light," Jim remembered. "He had developed a heated stage with a quartz slide coated with tin oxide. This kept the liquid crystal compounds liquid. The fact that the heated stage could be controlled by music and movies was fascinating."

Stürmer gave Jim the "best present" ever, the four books written by Otto Lehmann, the early pioneer in liquid crystal research, as well as a book by M. E. Huth, who published a paper, "Character of the Double-Refraction of Liquid Crystals" with D. Vorlander in 1911. Being able to discuss these books with a native German speaker helped Jim gain insights into the early observations of these liquid crystal pioneers.

He was fascinated by the optical properties of liquid crystals. "When I started looking at liquid crystals, their optical activity caught my eye," he remarked. "They were intriguing as I got more and more into them. I found all kinds of things people hadn't thought about. They were the opposite of a mirror in terms of polarized light. It was great fun."

Jim obtained two significant scientific results when investigating mixtures composed of cholesterol butyrate and cholesterol tetradecanoate. First, he determined that the change in color with temperature as a function of the mole percent of the component materials was a straight line when plotted on a log graph. (The mole is a unit of measurement used in chemistry to express the amount of molecules.)

Second, he found that the temperature of the clearing point also had a straight line relationship to the mole percent of the component materials. This meant that he was now able to determine the properties of the liquid crystal mixtures by varying the proportions of the constituent materials. He knew what color would appear at what temperature and could log that information on a graph. He was also able to change the clearing point, the point at which the liquid crystal materials turned from opaque to clear. Formerly liquid crystal materials would separate when he tried to blend them, but now the materials stayed blended.

INVENTING THE FIRST CHOLESTERIC OPTICAL TEMPERATURE MEASURING DEVICE

With these breakthrough results in hand, Jim began to design an optical device using cholesteric materials that could measure and visualize temperature. This would represent a breakthrough in being able to assess temperature simply by coating a surface and being able to "see" how hot or cold it is.

His initial goal was to create a device that could detect and show the infrared light emitted by natural objects. This was an extremely difficult design task because all objects with a temperature radiate infrared light. This includes not only objects of likely interest such as people and vehicles, but conventional components such as camera lenses, film and other types of detectors that might be used in the visualization apparatus itself.

Since the temperature would be measured over the entire surface of an object instead of at a few points on its surface, Jim had to figure out how to make its visualization be a continuous image, sort of like a satellite map of the ground below. In addition, the sensitivity requirement was very challenging. Slight temperature differences had to be made clearly visible.

The device also needed to have a rapid response time. If the temperature of the target changed, the device had to detect and display the change in near real time. Compounding these difficulties was that Westinghouse provided no funding to purchase components for the device or hire services.

Jim broke the overall task into individual problems and resolved each one at a time. His first problem was how to deposit a uniform, thin film of cholesteric liquid crystal on a support layer. The support material could not be soluble in the liquid crystal, and also had to be as thin as possible, since any support would add an additional heat load in series with the cholesteric thin film. Jim remembered a high school experiment he'd watched in which a thin film of cellulose was cast on a substrate by floating the cellulose on water. He used a variation of this approach to cast the cholesteric film.

For the support layer, he used a four micron thick Mylar film that he obtained for free from DuPont. He wasn't sure what to use to hold the film until it occurred to him that an embroidery hoop might suffice. He stretched the Mylar across the hoop, and deposited a perfect cholesteric thin film on the Mylar using a water cast process. Next, a means was needed to enhance the effectiveness of heat absorption into the Mylar substrate. Jim sprayed the rear surface of the Mylar with Krylon flat black paint. With this, he created a cholesteric liquid crystal-based imaging device that could be used to visualize thermal patterns.

None of Jim's senior managers had been keeping tabs on Jim during this project, but now Max Garbuny and Tom Vogl wanted Jim to meet with them and explain what he had been working on. When Jim started his presentation, Max got so excited he ran out of the room to get John Coltman. Jim did the presentation for John, who told Jim to concentrate on cholesteric liquid crystal compounds (which Jim had been doing anyway). Jim then demonstrated an interim apparatus he had developed to show the potential of his thermal visualization approach. It was an envelope from an x-ray intensifier tube fitted with a rock salt window which he had coated with a film of cholesterol nonanoate. Management declared the project official and asked Jim to build the device.

The final apparatus Jim built included the liquid crystal imaging device as well as optical and other components of a more conventional nature. The last issue that Jim had to address was a means of uniformly and precisely maintaining the cholesteric in its liquid crystal range. This was necessary because the extreme sensitivity of the color play in the cholesteric was subject to variations in room temperature and the input of heat from the image itself. Although it is a considerable simplification of the actual sophisticated apparatus,

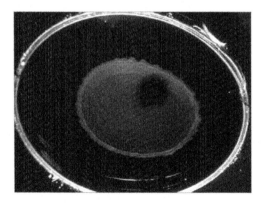

The imaging device utilizing cholesteric liquid crystal.
The darkest spot in the center upper right is Jim's fingerprint
made by touching the insulated table top, 1960

his solution was to use a heat lamp to illuminate the sample. A dimmer switch or variable transformer was used to control the heat lamp, which would precisely control the temperature.

The ultimate product of Jim's work was a device later named the Mesoscope. The world's first liquid-crystal based display device, it was so sensitive that its ability to discriminate temperature features was limited only by thermal noise. Not only had Jim met the goal of developing a highly innovative temperature measurement and visualization device based on state-of-the-art technology, but he had fabricated it at almost no cost through the use of inexpensive, commonly available components.

This powerful technique of measuring the thermal temperature of objects is still in use today. As Jim, Max and Tom Vogl wrote in the introduction to the patent application Westinghouse filed in 1960 for this device, "in many applications, it is desirable to provide means for giving a visual indication of the temperature of an object... A direct visual indication that an object is so hot that contact with it is dangerous would prevent many burns."

The patent described how thermal imaging could be done by directly placing the material on the object, or by viewing objects from the infrared radiation they emit. The challenge with the latter technique was "to convert a heat image which is invisible to the eye into an image which can be seen." While devices existed that did this conversion, they either required complex electrical circuitry, which made them costly, or were not sufficiently sensitive to distinguish between objects that emitted radiation at similar intensities.

Representatives from Fort Belvoir who were invited to see the Mesoscope contracted Westinghouse to develop it for practical use. With this government funding, Westinghouse and Jim were now officially in the liquid crystal business. Jim's device was truly a breakthrough in the field and was put to practical use by other Westinghouse scientists, who designed bolometers and thermometers that visually displayed the temperature of an object. The cholesteric temperature measuring device is "the most sensitive temperature-measuring device known to man," Jim later wrote. "Along with the unusual temperature sensitivity, the material is capable of very high resolution in the order of 1,000 lines per inch as well as time constants as short as 10 milliseconds. This would indicate that on a square inch of material, the temperature could be measured 100 times per second at 1 million points."

All of the liquid crystal research at Westinghouse was confidential and nothing about it was published for several years. An article Jim wrote on the Mesoscope in 1960 was the first public disclosure of the device.[9] Jim also wrote an article for the *American Journal of Physics* explaining how students could build this device as part of their laboratory course work.

Fort Belvoir was happy with the work Jim did on the Mesoscope and renewed the contract twice between 1959 and 1962. As part of the last contract, Westinghouse and Jim delivered hardware: an infrared image converter apparatus. By obtaining these contracts, Jim showed his ability to attract government funding. In the real world, this ability was as crucial as technical mastery and Jim had it in abundance.

After Westinghouse successfully completed the program with Fort Belvoir in 1962, the Westinghouse Defense and

Space Center made an appointment for Jim to meet with Harold Brown, a nuclear physicist who was then the director of defense research and engineering at the Pentagon. Brown, a previous director of the Lawrence Radiation Laboratory, later became secretary of the Air Force, president of the California Institute of Technology and, in the Carter administration, the secretary of defense. The purpose of the meeting was to make a wider circle of Department of Defense personnel aware of progress in the science and technology of liquid crystals. Although an informed and sophisticated man, Brown canceled the meeting. In an echo of the skepticism some of Otto Lehmann's peers in the scientific community expressed after Lehmann announced the existence of liquid crystals in 1889, Brown called Jim a fraud, declaring, "There is no such thing as a liquid crystal."

5. INCUBATION OF THE LCD AT WESTINGHOUSE

> *"An egg of a new technology had been found— now what remained was to hatch it and nurture it to adulthood. Not always an easy task."*
> —Jim Fergason

LARGE COMPANIES ARE constantly reorganizing, and Westinghouse was no exception. In late 1960, they formed an imaging group and named Jim principle investigator. In the span of just three years, Jim had gone from being a new hire fresh out of college to the head of the first group in the world doing industrial research into liquid crystals.

From 1960 to 1963, the imaging group studied the molecular structure, optical properties and technological possibilities of liquid crystals. This research helped clarify the unusual molecular architecture of liquid crystals as well as the "remarkable ability of liquid-crystal substances to register minute fluctuations in temperature, mechanical stress, electromagnetic radiation and chemical environment by changing their color," as Jim later wrote for *Scientific American*. The group conducted thermal measurement and visualization, researched imaging devices and developed television technologies. Contracts with various government agencies paid for all of the research costs as well as generating considerable profits for Westinghouse.

Jim with thermal imaging device and his handprint on the cover of *Iron Age*, 1966

JIM'S ASSISTANTS AT WESTINGHOUSE

Jim was assigned a skilled electrical technician named John Conroy. John was a great person and remained friends with the Fergasons for years.

John was helping Jim perform a chemical synthesis one day, making hard-to-find alcohols from commonly available acids. They were working with a hypergolic chemical called lithium aluminum hydride. Hypergolic chemicals are used as rocket fuel and when exposed to moisture, the chemical spontaneously combusts. About a gram of the chemical had been loaded into a flask that must have contained a trace of residual moisture. This caused a chemical reaction that began melting the flask. The two men ran into the parking lot with the flask, intending to pour the material onto an asbestos mat on the ground and let the chemical ignite and burn out. In their haste, they didn't notice that it had rained earlier in the day and the ground was damp.

Years later, Jim could still close his eyes and vividly see what transpired. As he leaned over and poured the chemical onto the mat, a tiny crystal rolled onto the damp ground and ignited, also igniting the mat. A flash of infrared light and thermal heat hit Jim in the face, causing first and second degree burns. The pen protector in his breast pocked melted. He was also burned on both arms, but luckily his eyes were protected by his glasses. Fortunately, John wasn't burned at all.

By the time Jim reached the nurse's station, he was going into shock from the pain. He demanded an injection for the shock, but the nurse washed him instead because she thought he'd been poisoned with lithium. At the hospital a blood type sample was taken from his toe because of the burns on his arms. He was in the hospital for two weeks. Blood samples continued to be taken from his feet. The burns on his face made it impossible for him to shave and he went home with a beard. After taking one look at their father, Jeff and Terri went running to their mother. A now favorite Fergason family photo was taken to commemorate Jim's new look before he shaved.

Jim with Terri and Jeff after the accident, 1961

John Conroy was promoted to engineering associate. Fred Davis, a creative thinker who had learned his way around a lab during his military service, replaced him. Newton Goldberg, a PhD in chemistry, was also assigned to Jim's project.

LCS COLOR CHANGING SPEED

One of Jim's first tasks was to determine how quickly a cholesteric liquid crystal film in a bolometric imager would change color. After much trial and error, he built an apparatus to test the speed of response. It had an anodized aluminum block as a heat sink and was coated with a thin polymer layer that served as a thermal insulator. He evaporated a nickel-chromium resistance film on the insulating layer. He was surprised at how fast the response time was when he coated the surface with a properly prepared liquid crystal. The liquid crystal changed color in milliseconds. This color sequence occurred faster than the eye's visual response.

Jim showed Max Garbuny the spectacular color switching. Max immediately brought over John Coltman for a look. John went and got Clarence Zener, of Zener diode fame, who was the director of research for Westinghouse. Zener asked if anyone else knew about the apparatus, and Jim replied that they were the first to see it. Zener told Jim not to talk about the device outside of the company.

One of the imaging group's main goals was to make a display based on the cholesteric phase. Within several months, Jim and his colleagues, including Art Andersen, the head of the laboratory imaging group, had a demonstration model based on a resistive heating array. Jim said that it was fascinating to watch the color switch as a variable current was applied. With increasing frequency, the color was no longer visible to the eye. The switching was still occurring, but so fast the eye couldn't see it.

The company built several different types of displays based on color change and cholesteric liquid crystals. Jim was able to build a matrix array and found driving the array to be quite easy. The power consumption was uncertain because of variations in the thermal environment. The temperature of the display had to be controlled in order to keep it within its operating range to prevent the color from changing, which occurred when the temperature changed by only one or two degrees.

The company filed patents for the imaging group's developments just before disclosure requirements made it necessary to do so, or just before undertaking a contract. Within three months of disclosure, the company paid a $50 bonus to the inventor. When it was determined that the disclosure was patentable, a second $50 bonus was issued. After the patent was issued it could be designated "meritorious." Jim had over 60 patent awards, including nine filings and one meritorious award. The extra money helped the Fergasons support their growing family, which included a third child, John David, born on December 27, 1962.

FIRST LIQUID CRYSTAL DISPLAY

In 1964, by building on the Mesoscope work and through Jim's efforts, Westinghouse and the imaging group were awarded a contract from the Rome Air Development Center, a U.S. Air Force lab, to develop flat panel display technologies. Westinghouse developed five different types of device configurations: an image intensifier, a nonlinear resistor, which was the forerunner of the active matrix, a transfluxor, a memory core and bistable cadmium selenide. All employed the same type of temperature-sensitive cholesteric liquid crystal film, but used different means to establish the thermal pattern. The performance specifications of the devices varied, but collectively the devices demonstrated the ability to display three or four colors on a black background. They also displayed from 104 to 106 picture elements, had response times from 0.2 seconds and resolutions of 10 lines/inch. As part of the contract deliverables, a 10" x 12" display screen was provided. The development of this liquid crystal-based technology showed that there was potential to produce a full color video display.

Jim created an invention to measure the wall thickness of a pipe thermally, using the ability of cholesteric liquid crystals to measure minute changes in temperature. He got the idea for the invention from drinking his morning coffee and noticing that the thickness of the mug determined how quickly he felt the heat of the hot liquid. The thicker the cup the longer it took for the heat to penetrate it. He applied this idea to a wall pipe. Previously, the wall thickness had to be measured mechanically, by cutting the pipe open, which destroyed the pipe. Jim's invention allowed "non-destructive" measurement of wall thickness using surface temperature.

Westinghouse introduced a commercial product based on this work, named Spectratherm. The product, a set of cholesteric liquid crystals that collectively covered the room temperature range, was the first of many based on liquid crystals that Westinghouse and other companies developed. *Industrial Research* magazine chose Spectratherm as one of the top 100 products of 1965 for its prestigious "100 awards." This award recognized Jim for helping Westinghouse develop the world's first commercial liquid crystal product. The product was also used by artists who collaborated with engineers from Bell Telephone Laboratories to develop technical systems used in a series of 1966 performances. For example, several of the dancers in *9 Evenings: Theater and Engineering* held at the 69th Regiment Armory in New York City performed wearing a coating of Spectratherm.

The liquid crystal project had met all its goals, but Jim determined that the thermal imaging device would not be suited for military applications because it could not withstand the rigors of the military environment at that time. (This has changed, as thermal imaging systems are widely used in the military, such as the FLIR thermal imaging infrared cameras.) Additionally, such devices wouldn't meet the needs of the broad-based display market. Nonetheless, the results of thermal imaging were innovative.

Jim realized that he needed to develop a display technology that was based on some other alternative to thermal activation. He knew that several researchers had investigated the effects of the application of external fields to liquid crystals. Much of this early work was inconclusive because the researchers had not determined whether the effects observed were caused by the material itself or impurities in the liquid crystals. The early researchers more often used magnetic rather than electric fields, because electric fields caused ion flow, which broke down the liquid crystal material.

Jim's new approach was to base the devices on electric field effects in cholesteric liquid crystals, rather than on a thermal effect. For this approach to succeed, he needed to prepare cholesteric liquid crystal materials of such surpassing purity that the resistivity of these materials would allow an electric field effect device to be fabricated and operated. He developed the means to purify cholesteric liquid crystal materials that had a resistivity on the order of 10^{16} ohm-cm, making the materials very high quality insulators.

All electrically insulating materials have a dielectric constant. This parameter indicates the magnitude of the charge that can be stored when the material is used to separate a pair of metal plates. Materials like polyester films are used to separate the plates in a capacitor because they have a very high dielectric constant. The properties that liquid crystal materials exhibit in response to an electric field are called dielectric properties.

Liquid crystal structures respond to an externally applied electric field in one of two ways depending on their molecular structure. In the first type, the liquid crystal molecules have a property called a positive dielectric anisotropy. The long axes of liquid crystal molecules of this type align parallel to the direction of an externally applied electric field. The second type of liquid crystal molecules have a negative dielectric anisotropy, and align with their long axes perpendicular to an externally applied electric field. The polarity of the liquid crystal determines in which of the two directions the liquid crystal molecules will align themselves in an electric field.

Jim conducted some initial experimental exploratory work and found that cholesteryl alkyl carbonates were not visually affected by an electric field. Cholesteryl chloride was affected, but the pattern of response to the applied electric field did not comply with the simple picture discussed above. These observations were the starting point for Jim's new line of research, an approach based on phenomena that had not previously been recorded. These materials had low dielectric coefficients and an extremely high resistivity.

Jim said, "One of the big advantages to working at a research laboratory such as Westinghouse was the opportunity to learn unpublished tricks of the trade. Cholesteryl derivatives were very good insulators, but in the liquid phase they still have some percentage of ions. It was very difficult to remove those ions. However, I was friends with the chemists in the insulation department, who I discovered were using Fuller's earth or Floridian clay to absorb ions. I remembered my childhood experiences making sorghum molasses using Fuller's earth to clarify the juice from the cane. It struck me that I could do the same thing with liquid crystals. I mixed the cholesteryl liquid crystals with spectrographic grade pentane. Then I made a slurry with the thermally activated Fuller's earth, filtered it with a Millipore filter and evaporated the pentane. The result was a between three and four orders of magnitude increase in resistivity to greater than 10^{15} ohm-cm. This material was used to make the first field effect liquid crystal displays."

Fuller's earth is also known as "kitty litter." Jim had used it to increase the resistivity of his materials. He remarked, "It occurred to me that taking an experimental fact as a partial motivator to take a further experimental risk in another direction was the best of science. I used the clay in an innovative way in an entirely different application."

Jim investigated an approach in which the cholesteric liquid crystal display was activated by charge deposition. The photo on page 67 shows the apparatus that was constructed on these new principles. The 110 volts are applied through use of a variable transformer. The top electrode is tap water and the bottom one a coating of flat black spray paint.

To construct the device, Jim developed a process to apply an even, thin layer of liquid crystal between two flexible Mylar substrates. The liquid crystal was dissolved in a volatile solvent and then sprayed onto the lower substrate using an artist's airbrush. An edge of the upper substrate was put into contact with the liquid. The weight of the upper

Cholesteric field effect device

substrate in conjunction with the surface tension of the liquid caused the top substrate to slowly contact the liquid on the lower substrate. The entire gap between the two substrates was gradually filled with a thin layer of liquid crystal. This procedure was performed in a vacuum to exclude air. When Jim invented the twisted nematic liquid crystal display years later, he adapted this process to fill the substrates of the TN-LCD with liquid crystal material.

The device was constructed under a research program investigating liquid crystal media for electron beam recording, funded under a contract with the Air Force Avionics Lab, Air System Command at Wright-Patterson Air Force Base. The result was the first practical application of an operating electric field effect liquid crystal display device.

"Here was another key step in the development of liquid crystal displays," Jim said. "Two ways were found to drive a cholesteric display. The most common is to drive the display with a low impedance source which untwists the structure, and with the untwisting of the structure, the phase changes from a cholesteric to a nematic." This device was patented.[10]

The final report for the Avionics Lab contract was published after Jim left Westinghouse. Much of the work undertaken during this period was documented and later bolstered by a patent application filed in November, 1963. It was the first patent ever issued embodying an electric field effect in a cholesteric liquid crystal device. "This was a step toward modern display systems, which were the forerunner of things to come," Jim wrote in an article about this work. "Up until this time we were using liquid crystals to display existing thermal patterns, but here was a point I realized we could generate those patterns in real time, using electronic inputs, that were fast and easy to read."

Jim often talked about liquid crystals with Westinghouse co-worker Peter Brody, who worked in another division but had an office next to Jim's. Brody's project related to the use of thin film transistors to make an active matrix array to drive pixel elements in display applications. Some of Brody's work utilized the array in conjunction with an electroluminescent display. Brody left Westinghouse in 1979 to form a company called Panelvision. Capitalizing on experience gained in part at Westinghouse, he switched from electroluminescent materials to liquid crystals. Panelvision was the first thin-film transistor liquid crystal display developer and manufacturer in the world. Brody and Jim stayed in touch.

EARLY COMPETITORS IN LC RESEARCH

RCA Laboratories in Princeton, New Jersey was also conducting liquid crystal research. In the 1950s, RCA general manager David Sarnoff reportedly told his scientists to "create a TV you could hang on the wall." Richard Williams at RCA Laboratories began experimenting with liquid crystals in 1962, following a line of inquiry that had roots in work conducted in the early 1900s by Vsevolod Freedericksz at the Optical Institute in Petrograd. This early research investigated the effects of competition between surface alignment forces and an applied field. One of the most important observations Freedericksz made was that the threshold value of a field must be achieved before an optical effect occurs. This threshold effect is crucial in the design of the electronic drivers used in modern liquid crystal displays.

Freedericksz was probably the first to observe the appearance of periodic hydrodynamic domains when an electric field was applied to certain types of aligned nematic liquid crystal materials. Richard Williams applied a voltage to thin layers of liquid crystals and observed that a striped pattern resulted. He initiated the idea of using the hydrodynamic instability forming in the liquid crystals, an effect later called "Williams Domains," to make electrically operated displays. Williams also concluded that testing this idea and making displays would be too big a project to undertake.

However, Williams' colleague, George H. Heilmeier, became interested in making displays. Heilmeier was then a PhD student at Princeton working part-time at RCA. After Heilmeier received his PhD in 1964, Williams went on sabbatical, and Heilmeier replaced him. In 1964, he and Louis Zanoni built a device popularly called a dynamic scattering display. Later, RCA developed guest-host displays, a type of dynamic scattering display in which the active component is composed of dichroic dye (called the guest) dissolved in a nematic liquid crystal (called the host). With his synthesis of liquid crystal compounds at Westinghouse, Jim had laid the groundwork that enabled RCA to begin making displays.

POPULAR SCIENCE MAGAZINES TAKE NOTICE

In 1964, *Science* magazine asked Jim to write an article on liquid crystals. He felt the article he wrote was excellent, but a peer reviewer demanded extensive revisions nonetheless. Jim considered the changes unnecessary but complied.

He also submitted a requested article on liquid crystals to *Scientific American* in which he summarized research, explained the basic science and discussed current and potential applications. The article appeared in the August, 1964 issue. Reader response was overwhelming. Over 370,000 copies of the magazine were sold and Jim received thousands of letters about the article, an astonishing number considering that liquid crystals were virtually unknown. The popularity was such that *Science* magazine canceled the article they had requested Jim revise, saying they had really wanted an article like the one he wrote for *Scientific American*.

"It took from 1957 until really 1964 for the potential of liquid crystal products to be recognized enough to get real interest and funding," Jim told science reporter Linda Hamilton in 2002.[11] "That was when I published the article in *Scientific American*. It was a big leap forward."

Jim had vaulted onto the international stage as one of the foremost experts in liquid crystal science and an innovator in the field. Companies realized that liquid crystal products could potentially be profitable and began funding research. Many visitors came to Westinghouse to discuss liquid crystals with Jim.

Among them was Dr. Bruno Hampel from E. Merck, Inc. After the visit, Merck decided to sell a product line of liquid crystal materials for the first time in many years.[12] Donald C. Batesky from Eastman Fine Chemicals (Eastern Distillation) also visited Jim. Kodak published a periodic newsletter and the cover illustration of one showed Batesky reading a copy of Jim's *Scientific American* article on an airplane. An ad inside the newsletter announced the availability of liquid crystal materials from Kodak. Kodak also ran the ad in *Chemical Week*, *Chemical & Engineering News*, *Physics Today*, *Medical World News* and many other periodicals.

Contact with Jim also seemed to have influenced many other companies. After Wayne Woodmansee of Boeing visited Jim, Boeing embarked on a non-destructive testing program using liquid crystal materials and received a patent.[13] The National Cash Register company undertook a program related to the encapsulation of cholesteric liquid crystals, a business that continues to this day, and received a patent.[14] Hoffmann-La Roche began selling liquid crystal materials and included an article in their company newsletter on cholesteric liquid crystals that mentioned Jim. Chevron and Xerox also began liquid crystal research and development programs soon after publication of the *Scientific American* article.

WESTINGHOUSE PROBLEMS

During this time, Westinghouse president Donald Burnham and vice president Bill Shouppe took the president of

Montgomery-Ward, Robert Brooker, on a tour of Westinghouse. Brooker told Jim he had read the *Scientific American* article and had never met such a distinguished scientist and author before. Laying it on a little thick, he asked for Jim's autograph. During this exchange, Jim heard Burnham ask Shouppe why the imaging group was doing research and development into cholesteric liquid crystals. For some reason Shouppe replied, "I don't know." This reply concerned Jim. Years later, he told a reporter that Shouppe could not give Burnham a "profit-related answer" and that Westinghouse wasn't interested in using liquid crystals for temperature control.

Like many large companies, Westinghouse decided which discretionary programs were to be funded during the upcoming year. Shouppe adopted a "ladder" approach to funding in 1965, prioritizing all of the programs competing for funds from the top rung down. He ranked the cholesteric research and development program below the funding cut-off point. Liquid crystal-related activities at Westinghouse would still continue through government-funded contracts, but if Jim wanted to sustain his program, he would have to scrounge for support. He was fed up and began looking for a new job.

One of his last visitors at Westinghouse, physicist and liquid crystal researcher Alfred Saupe from the University of Freiberg, was someone Jim had long wanted to meet. Alfred had written his 1958 doctoral dissertation, under the supervision of Wilhelm Maier, on the development of a microscopic mean field theory for liquid crystals. Called the Maier-Saupe theory, it is widely accepted as the standard description of liquid crystalline order. Jim had been reading Alfred's papers for eight years and agreed with his findings.

The two hit it off. They discussed the unusual properties exhibited by cholesterol nonyl carbonate, a cholesteric liquid crystal material. The peculiar aspects of this material included it being a highly viscous room-temperature liquid crystal that did not crystallize and had no double refraction (birefringence), that is, the refracting of a ray of light into two separate rays. It is an optical phenomenon characteristic of all solid crystals and most liquid crystal substances. Cholesterol nonyl carbonate had a high degree of optical activity and exhibited a "blue haze."

Alfred reported to Jim that this was the mysterious "blue phase." Until about 2004, blue phase materials existed as somewhat of a laboratory curiosity.

THE 1965 LIQUID CRYSTAL CONFERENCE AND THE MOVE TO KENT STATE

In July 1965, Jim presented his findings on cholesteric liquid crystals at an international conference on liquid crystals that he helped organize with the director of the new Liquid Crystal Institute at Kent State University, Glenn H. Brown. A chemist and Ohio native, Brown had first stumbled upon mention of liquid crystals when he was a professor at the University of Cincinnati in the 1950s and was looking for an intriguing research topic for his graduate students.

Glenn had written a 1957 literature review summarizing previous publications on liquid crystals that appeared in *Chemical Reviews,* but Jim did not regularly read chemical literature at the time. He first heard of Brown from a semiconductor scientist, Henry Levenstein, who consulted at

Westinghouse. Levenstein had become interested in liquid crystals after meeting Brown on a business trip. Through Max Garbuny, Levenstein learned there was a liquid crystal development program at Westinghouse and introduced Brown to the company's liquid crystal group.

Brown's literature review is now considered historically significant, but at the time it did not inspire much interest in liquid crystals. There was so little interest in the topic that an earlier attempt Brown had made to organize a liquid crystal conference failed. Jim's *Scientific American* article appeared in a popular journal rather than an academic tome, and its emphasis on practical applications spurred so much interest in the field that Brown, with Jim's help, was now able to get a conference off the ground.

Jim presented two papers: "Cholesteric Structure—I Optical Properties," and "Cholesteric Structure—II Chemical Significance" (which he co-presented with N.N. Goldberg and R.J. Nadalin). Both were well received by all of the scientists present, except those from RCA who had an attitude that they were major leaguers in liquid crystal technology compared to all others. They criticized one of Jim's papers for being highly experimental rather than a more "respectable" theoretical study, although the paper had a complete analysis based on standing waves in a twisted structure. Nevertheless, Jim's papers sparked additional research and led Dwight Berreman and Saul Meiboom to enter the field at Bell Labs. His papers are still cited in patent applications today and were published in *Molecular Crystals*.

Jim made no secret that he was looking for a new job and many employers were interested in hiring this star in the field.

He interviewed with Hoffmann-La Roche, Merck, the Defense Intelligence Agency and Glenn Brown at Kent State. Glenn enthusiastically courted Jim, whom he wanted to hire as associate director to help him develop the Liquid Crystal Institute into an international leader in liquid crystal research. Glenn also wanted to attract business to LCI and Jim had industry connections. While Glenn couldn't match the salary offers from companies, he offered Jim job flexibility and independence, including the opportunity to pursue independent consulting. The job description was a tempting, "do whatever you want."

Jim accepted the offer and as a formality, Glenn requested letters of reference from Westinghouse. Max Garbuny wrote in his reference letter:

> I have known Mr. Fergason since he joined my Optical Physics Section almost ten years ago. It quickly became apparent to me that Mr. Fergason possessed an unusual degree of scientific skill and insight—perhaps one should call it intuition—all the more astonishing since he joined our group with the rank of a bachelor's degree. This ability showed itself especially in the understanding of the physical properties to be expected from a given or projected molecular structure. I might mention here that his capability of synthesizing organic compounds represented in itself no small contribution, since some chemists of long experience had claimed in advance that such substances could not exist. I assigned him to the study of cholesteric compounds after his tour of duty with the army, and the results of his work are, of course, well known to you.[15]

He tempered this glowing praise in the next paragraph. "Jim has overcome rather well an early difficulty of communicating his ideas. I believe, therefore, that he will be doing rather well [sic] as a lecturer." Given that Max once tossed a cigarette into a container of liquid oxygen, what he wrote next would have given Jim a good laugh if he ever read it: "The thing to be watched, perhaps, is a tendency to procrastinate, and, occasionally, to be a little careless with experimentation. However, to the extent to which these habits may have constituted drawbacks, he certainly has very much improved in that respect."

Westinghouse tried to convince Jim to stay, but to no avail. On his last day of work, June 1, 1966, he met with someone from human resources to sign documents. The HR person then handed him a check for the amount of his contributions to the Westinghouse employee's pension plan. Jim asked why the check didn't include Westinghouse's ten years of matching contributions, and was flummoxed by the answer. The company contributions vested after ten years and Jim had been at the company for nine years, eleven months and two weeks. If HR had told him this, he would have delayed his departure by two weeks. The HR person had the gall to remark, "What a shame. If you had only stayed a few more days, all that money would have been yours."

Westinghouse's liquid crystal program dried up by 1969 and the company was out of the liquid crystal business. There were many reasons for this, but one was that the technology had lost its most effective champion within the company, Jim Fergason.

6. THE LIQUID CRYSTAL INSTITUTE

> *"Jim is a very important person in the success of our Institute."*
> —Glenn Brown

THE LIQUID CRYSTAL INSTITUTE (LCI) at Kent State University was in its infancy when Jim started working there and, as Glenn Brown's early correspondence reveals, he valued the ability of his new associate director to attract publicity. For LCI to thrive, it needed a public profile.

The summer before Jim began his new position at LCI, the University of California in Los Angeles had invited him to teach a short course on liquid crystal science. The course organizers wanted a co-teacher with a PhD so Jim invited his new boss, Glenn Brown. Jim organized the course and taught most of the topics. Having the director and associate director of the new Liquid Crystal Institute leading the class was great publicity for LCI.

"Liquid Crystals: Their Physics, Chemistry and Uses" ran from June 20–July 1, 1966. On the weekends, the Fergasons took their three children to Disneyland, Knott's Berry Farm and Marine World. After the course ended, the family returned to Pittsburgh to pack up their old home.

Jim made several trips between Pittsburgh and Kent moving the family's belongings. On one of these trips, Jim arrived in the evening and Terri's turtle Myrtle had escaped her tank and disappeared. Not wanting his daughter to be upset, he went over every inch of the car looking for her. He pried out the back seats and found her alive behind one of them.

The move from industrial Pittsburgh to a small college town was a welcome change for the family. Dora enjoyed the rural feel of Kent and the children enjoyed the open spaces. Housing prices were higher in Kent than Jim and Dora had anticipated, and Dora's mother had to loan them the money to buy a roomy brick house conveniently located one mile from campus, in a neighborhood where other KSU professors lived. The house sat on an acre of land, which allowed Jim to reconnect with his Missouri roots by planting corn, potatoes, beans, carrots and tomatoes. He also planted fruit trees and Dora cultivated a backyard flower garden. Their sons Jeff and John played football in the large front yard and Terri searched the field bordering the back yard for rabbits.

GETTING LCI OFF THE GROUND

Jim had been instrumental in reviving interest in liquid crystals, and in hiring him as the associate director of LCI, Glenn Brown had scored a coup. The two men shared a keen interest in liquid crystals, but were opposites in other ways. Glenn was 19 years older than Jim, a reserved, rather inscrutable person. A career academic with a PhD in chemistry, he was a savvy negotiator of academic politics and a master networker. While he was passionate about liquid crystals, he was not a groundbreaker in the field. His 1957 publication was a review of the literature on the subject and he was still a proponent of the outmoded swarm theory.

In contrast, Jim was dynamic and open-minded. His breakthroughs in the liquid crystal field had come about through hands-on experimentation. This was fueled by what his Westinghouse boss Max Garbuny called his scientific intuition. Ken Marshall, a chemist who worked at LCI under both men, remarked to us that, "Glenn was an academic, and Jim was a practitioner and somewhat of a maverick."

Jim was excited about his new job, but had concerns when he entered the Institute's labs for the first time. They were empty except for a fume hood and laboratory benches. Although Jim's salary was assured, there was little money for purchasing equipment. Undeterred, Jim began setting up his ground floor office and lab. He bought a Leitz polarizing microscope with a heated stage, not dissimilar to the one Otto Lehmann used in his pioneering liquid crystal research, and a DuPont Differential Thermal Analyzer. With these instruments, he planned to measure the optical and thermal properties of liquid crystal materials composed of molecules with specifically designed properties.

LCI was housed in the Lincoln Building, a cheaply constructed building on Lincoln Street which the university had leased from a real estate developer named Daniel C. Jones.

Jones also had a law degree and undergraduate and graduate degrees in engineering, and worked as an assistant to Glenn Brown.[16] The local newspapers picked up on this flagrant conflict of interest, and the attorney general for the state of Ohio initiated an investigation of Jones. To end the investigation, he donated the building to the state. LCI shared this unprepossessing building with researchers from other university departments and were always cramped for space.

Jim soon discovered that Glenn didn't spend much time at the Lincoln Building. The university had hired him in 1960 to be the chairman of the chemistry department. When he became dean of research in 1963, he gave up the position of chemistry chair, but was jockeying to get the chairmanship back and was preoccupied with chemistry department politics. He therefore worked from the dean's office, located in a separate building. So did LCI's secretary. This meant that whenever Jim needed secretarial help, he had to trot over to the dean's office.

LCI gradually acquired more staff. First was Adriaan de Vries, a scientist and expert in x-ray technology from the Netherlands. When Adriaan entered his lab in the Lincoln building, he found it as empty as Jim's. He ordered his most critical equipment, a Picker Corporation industrial x-ray machine, but then did almost no work during the nine months it took for the machine to arrive. Glenn also hired a young physicist, Ted Taylor, for a two-year assignment working with Adriaan in the x-ray study program. Ted was on a post doc appointment in the KSU chemistry department. He had grown up only 108 miles from Jim's family home in Missouri and had also attended Missouri University, but the two men had never met or heard of each other. Graduate students were also brought in to help out.

It was still months before the Picker x-ray machine would be delivered and Ted Taylor could assume his regular assignment with Adriaan, so Jim arranged for Ted to work with him. Jim hired Sardari Arora, a synthetic organic chemist born in 1929 in Lahore, British India. Sardari had survived the bloody partition of India into Pakistan and India in 1947. He was offered a post-doctoral position in the KSU chemistry department and moved with his family to the U.S. in 1964. The funding for Sardari's research ran out and he was happy to get the job at LCI, even though he knew the position wouldn't offer much range for independent creativity. He synthesized liquid crystal compounds that Jim had been designed specifically for their fluidity, dielectric and optical properties.

When Adriaan's x-ray machine finally arrived, he used it to execute a research project of Glenn's. The purpose of the project was to measure the structure of swarms in liquid crystal materials. However, the Elastic Continuum theory proposed by Hans Zocher in 1933 was supplanting swarm theory and from his work in cholesteric liquid crystals, Jim also knew that swarm theory was invalid and had stated so in his *Scientific American* article.

But Glenn was still wedded to swarm theory and told Jim that he had discovered a new nematic material that exhibited intense scattering of light at different wavelengths. He believed this provided evidence for the existence of swarms and asked Jim to examine the material. Using his new polarizing microscope, Jim found no evidence indicating the existence of swarms. The sample was clearly composed of two insoluble components. It was a perfect example of what the famous British biologist Thomas Huxley had said, "The great tragedy of science: the slaying of a beautiful hypothesis by an ugly fact."

Jim had to deliver the ugly fact to Glenn, who did not take it well. Nonetheless, the relationship between the two men remained harmonious for a while. Jim was a powerhouse of publicity for LCI. During his first year alone, he gave 13 lectures nationally on liquid crystals and their applications, while Glenn gave just three. Jim lectured at the NY Academy of Sciences on the "Application of Liquid Crystals in Medical Research," at Carnegie Tech on "Engineering Applications of Liquid Crystals," at an American Oil Chemists Association meeting in New Orleans on "Liquid Crystals and Living Systems" and at the University of Minnesota on "Recent Advances in Liquid Crystals."[17] He played a huge role in establishing LCI as a world-renowned center of liquid crystal research.

Dora also played a role in contributing to the institute's success, offering to host parties in her home because Glenn and his wife did not enjoy entertaining or drinking alcohol. Dora attended the university women's meetings. At one of these meetings, the wife of a university professor jealously remarked, "You live in such a large house. That's because your husband came from industry."

DEVELOPING BUSINESS FOR LCI

Initially Jim, Glenn and the LCI secretary were the only three employees funded by the university, but by the end of 1967, LCI was self-supporting, having attracted government contracts from Wright Patterson Air Force Base, the Night Vision Research Laboratory at Ft. Belvoir, the Advanced Research Projects Agency of the Department of Defense, and the National Institutes of Health. Jim hired new employees as part of a larger plan to mold LCI into an interdisciplinary center where both basic and applied research could be performed. He recruited professors and students from KSU's physics, chemistry and biology departments. In the October 3, 1967 issue of *The Daily Kent Stater*, he remarked, "With the continuing efforts of the research team, KSU may be noted in the future as the home of the liquid crystal research center of the world."

For work on the government contracts, Jim researched the use of cholesteric liquid crystals in thermal mapping applications and the detection and identification of organic vapors. He also visualized the thermal pattern of an air foil in a hypersonic wind tunnel and measured the thermal load on heat exchangers and electrical circuits. Additionally, LCI had a program with the Air Force to evaluate the practicality of using liquid crystals as sensors for analyzing pollution in closed atmospheres.

LCI landed a $50,000 contract to develop cholesteric materials for Hoffmann-La Roche to use in developing diagnostic instruments. Through this contract, Jim befriended some of the Roche employees at their U.S. office in New Jersey. These friendships would help him years later when he was engaged in a bitter patent battle with Hoffmann-La Roche, as will be presented later in this book.

From 1966 to 1967, Jim collaborated with a few medical schools on cholesteric-related research. He conducted a program on medical and biological thermal mapping with the KSU biology department and Portage County Hospital. Although this collaboration was new, Jim's idea that liquid crystals could be used for medical and biological thermal mapping was not, as he had previously conducted studies with two dermatologists at the University of Southern California School of Medicine proving that liquid crystal films could be used to identify skin diseases by thermal mapping. Based on the experience he had gained at Westinghouse designing an optical device that could measure and visualize temperature, he applied the idea to visualizing the temperature of the skin and used embroidery hoops to make portable liquid crystal films for clinical use. He also used the "Spectratherm," the set of cholesteric liquid crystals that Westinghouse had developed to measure surface temperature, in these studies.

A PhD candidate in physiology named Tom Davison participated in these thermal mapping programs and wrote his thesis on the medical and biological applications of liquid crystals. Tom's thesis advisor, Keith Ewing, also worked on the program, as did another graduate student named Max Chapman. LCI teamed up with doctors at the University of Southern California and the Roswell Park Cancer Institute in New York to study the medical applications of cholesteric-based thermal mapping. Jim's idea was to use cholesteric liquid crystals to create a visible, continuous thermal map on the skin surface of patients, an idea he'd successfully used in the dermatological studies. Doctors would then learn to interpret the pattern and use it to diagnose the medical condition of underlying structures.

Then the thermal mapping group needed to determine what medical conditions could be detected with a surface temperature mapping technique. They chose breast tumors because these are usually relatively close to the surface of the skin. A tumor is a rapidly growing tissue with a high concentration of blood vessels that slightly raise the local temperature. These two factors made it likely that breast tumors could be located using surface thermal mapping.

Developing a means to visualize the thermal patterns produced on the skin by use of a cholesteric was a technological challenge. Jim applied the same idea he had for the skin disease studies, for which he had created a high contrast background for the cholesteric liquid crystal by blackening the skin with a neutral coating composed of a low hydrolysis polyvinyl alcohol mixed with carbon black.

Tom Davison then recruited four female volunteers for the initial tests from among his close female friends and the girlfriends of his graduate school friends. Two female technicians performed, or were present, for all of the examinations. For the last study on breast thermography, Davison had to find ten co-ed volunteers who would agree to be examined daily for up to 45 days. The study was to determine the effects of daily activity, alcohol consumption, sleep habits, sexual activity and cigarette smoking on thermographic patterns and to assess whether these factors might affect accurate interpretation. Davison worried it might be difficult to recruit women who would be willing to have their chests painted black and covered with liquid crystals, but this was a college campus at the peak of the psychedelic 1960s and women volunteered in droves. They wanted to participate in a meaningful project and thought it would be fun to have their skin painted with liquid crystals.

The KSU art department read about the breast cancer studies and called asking for help putting on an art show using liquid crystals. Jim referred the call to Davison. "Members of the art department asked me to 'paint' really gorgeous, almost naked ladies with liquid crystals for an art show," Davison later wrote.

To Jim's dismay, Glenn Brown was against this experimentation and publicly stated that the use of liquid crystals to detect breast cancer might have potentially harmful effects. It was the first disparaging comment that Glenn had made about a program headed by his associate director and was perhaps an omen of what was to come between the two men. Glenn's comment also proved incorrect. In September, 1969, Jim was awarded $5000 by the Ohio division of the American Cancer Society, for pilot research on breast cancer thermography. The local Kent newspaper, *The Record Courier Review* reported that Jim was awarded the grant after the review group obtained assurance that liquid crystals are a safe material from the U.S. Public Health Service.

Jim forged ahead. The International Ladies Garment Workers Union adopted a liquid crystal-based breast cancer screening program, using a van to travel around offering screenings. The procedure was also used in France and Spain. At an international meeting on thermography, a U.S. participant stated that the use of liquid crystals might work for third world countries, but not for the U.S. The next speaker, who was from France, wryly remarked, "I guess France is a third world nation."

Meanwhile, Jim and the exercise physiology department conducted another thermography study, in which they visualized the dilation and constriction of blood vessels during exercise. The results were published, but without Jim listed as an author. This promising research program was never completed to his satisfaction.

While Jim was at LCI and after he left, he authored or co-authored 20 scientific articles on the possible biological applications of liquid crystals, including three related to the use of liquid crystal-based thermography to detect breast cancer. An article Jim co-authored with Tom Davison was published in the top medical journal *Cancer*, which was quite an achievement at the time. Davison was especially thrilled because he was just a graduate student. This publication drew extensive recognition for the thermography program at LCI. Davison and Jim also published a paper in *Obstetrics & Gynecology*, which was as prestigious a journal as *Cancer*. By this time, Glenn Brown had gone from criticizing the program to endorsing it.

LIFE MAGAZINE MAKES LIQUID CRYSTALS, AND JIM, FAMOUS

In response to increasing curiosity about liquid crystals, Glenn and Jim co-authored an article on the science of liquid crystals published by the *Journal of the American Oil Chemists' Society* in 1968. Glenn covered the lyotropic, or soap-like liquids formed by aqueous systems and Jim covered the physical properties that could occur in the structures associated with living systems. This article became one of the more popular Jim published during his long career, but it was a *Life* magazine article that made Jim's name known to the general public.

When Jim and Glenn were co-teaching a second short UCLA course in mid-August of 1967, a *Life* magazine photojournalist named Henry Groskinsky came to the UCLA campus to interview the two for a feature article on liquid crystals. Jim was again doing the bulk of the teaching and Groskinsky only had the chance to speak with him briefly, but he interviewed Glenn and spoke with most of the 11 course participants. They were scientists and researchers at Hoffmann-La Roche, Mobil Research and Development Corporation, Shell, Dow Chemical Company, Xerox Corporation and Wright Patterson Air Force Base. He also interviewed Jim's collaborators on ongoing LCI medical projects, including William Leonard, Gordon Stewart and John T. Crissey.

That December, Groskinsky called Jim and asked for permission to use his name in the article. Jim said yes. Glenn, ever the savvy promoter, wrote to the UCLA course organizers to let them know that the popular magazine would be carrying a feature on liquid crystals in its upcoming issue. He stated that he and Jim "knew nothing about the written portion," but that "we hope that UCLA and its support of the field of liquid crystals through its extension program would be mentioned" (it was not).

Life hit the newsstands on January 12, 1968. It sold for 35 cents. The movie "Bonnie and Clyde" had been released in 1967 with actress Faye Dunaway playing the part of the outlaw Bonnie Parker, and *Life* had a photograph of Dunaway, in costume, on the cover. But a photograph in the article on liquid crystals was so dramatic it stole the show. Now famous in liquid crystal circles, the photo showed a woman's bare back (the model was KSU art student Marci Murray) vividly colored in green, blue, orange and black. Murray's back had been coated with blackening material and then cholesteric liquid crystals. The vivid color patterns on her skin surface indicated the temperatures of the underlying tissue.

Jim's personal photograph of model used in the *Life* magazine article

Groskinsky also included an illustration of a large, liquid crystal-based signage display developed by Xerox but built along the lines of technology Jim had pioneered at

Westinghouse. This display development preceded that of the dynamic scattering mode display work that was later announced by RCA.

The article had the catchy title "The Chameleon Chemical: an oddity called liquid crystals proves to be of colorful use to doctors and scientists." It began, "James L. Fergason, associate director of KSU's Liquid Crystal Institute, is credited with the first practical use of cholesteric liquids for temperature measurement and the first use of cholesteric liquid crystal analysis and information display." The entire article was about Jim's research. Glenn Brown was not mentioned once.

Life magazine was then the most popular photographic magazine in the U.S., with millions of readers. Groskinsky had not told Jim that he was going to feature him as the lead liquid crystal researcher at LCI, and it seemed to be a major accomplishment for both Jim and LCI. In a letter dated January 15, 1968, Glenn wrote the following about Jim to Kent State University President Robert White, "I shall comment on personnel first. The salary adjustment for Jim Fergason is based on his fine contributions to Kent State. Our annual report reflects his attractiveness as a speaker on the subject of liquid crystals. He is a consultant to a couple of industries and receives many telephone calls from governmental and industrial laboratories seeking his counsel on research in the field of liquid crystals." The letter continued, "As you recall, he [Jim] received mention in the January 12 issue of *Life* magazine. Jim is a very important person in the success of our Institute."

LCI was Glenn's baby and publicity for it in such a popular magazine gave Glenn leverage to request more funding. He wanted to hire a second senior staff member to replace several part-time personnel and a third staff member, writing that, "As I have stated many times, we have a unique opportunity to be international leaders in the field...It is safe to write, I believe, that the Institute has brought the University more favorable recognition than any other activity on our campus."

Beneath the surface, however, Glenn was apparently upset that Jim had upstaged him in the *Life* magazine article. Jim later said that Glenn went "ballistic" about the omission of his name and seemed to blame Jim for it. He stopped talking to his associate director, communicating instead through endless memos. After a routine change in the lock to the Lincoln Building, Jim had to ask for a new key many times before one was issued.

Meanwhile, an investment banker and venture capitalist named Jim Bell read a reprint of the *Life* article in the *International Herald Tribune* while he was on a business trip to Amsterdam. Intrigued by the commercial prospects of liquid crystals, Bell called up Jim Fergason and proposed they go into business. Jim declined the offer, but the two men hit it off. They stayed in touch and later worked together on numerous projects for over three decades.

7. GREAT FRIENDS

"Jim is an extremely agreeable co-worker, very friendly, cheerful, and equally courteous to those in lower or higher positions than his."
—Max Garbuny, in a letter to Glenn H. Brown

DESPITE THE GROWING tension between Glenn and Jim, LCI was flourishing, with labs set up, more staff hired and new contracts and research projects under way. One of Jim's new hires was Tom Harsch, a young electronics technician who had worked as a research technician on the development of the next generation underwater acoustic homing torpedo. Both men had grown up in households where machinery was repaired instead of replaced, and both loved physics, particularly optics. Tom had been reading *Scientific American* since he was a teen and had read Jim's article on liquid crystals. Another commonality was that Tom had come to LCI from an industrial research facility rather than academia and had exposure to industrial values and business management practices.

In his first months at LCI, Tom grew restless. He had been hired to maintain the electronic equipment, and design and fabricate special laboratory instruments, but found he had nothing much to do. "Dr. Brown merely expected me to maintain the equipment at the Institute, offer my services to the researchers and do occasional administrative tasks for him," Tom recalled. In his previous position, he had executed increasingly challenging projects that allowed him to learn new skills, but the pace at the institute was much slower. That was before he started working with Jim.

Tom and Jacky Harsch, 1967

"Jim had plenty of ideas, was directing a number of programs, and seemed to welcome my help," Tom said. "He was young, smart and enthusiastic. He did meaningful work and was always generating new projects." Tom's first project for Jim was to design and build a temperature control mechanism for Jim's Leitz microscope. Thereafter, Tom conducted most of Jim's experiments and soon observed that, "Jim left open opportunity for unexpected results, rather than only confirming what he believed. His management style was to give me new ideas or material, discuss what he was looking for and leave me to devise some way that it might be observed and verified. I was learning again."

Jim explained his new ideas so thoroughly that Tom was able to build liquid crystal devices, assemble instruments and perform experiments without Jim standing over his shoulder. The experiments involved a lot of time-consuming trial and error before they worked. Whenever Tom was able to show something interesting, he called Jim. This method of working together saved Jim a lot of wasted time fiddling in the lab. It was also great for Tom because he learned a lot about liquid crystals and was excited to participate in advancing the science. Jim enjoyed collaborating with Tom and the two developed a lifelong friendship.

Jim was egalitarian, treating everyone he worked with as colleagues. He fostered an atmosphere that encouraged curiosity and was always happy to entertain discussions, particularly about liquid crystals. He was like the hub of a large number of spokes, with the spokes being the people he worked with. He created a universe of collaborators, always explaining the fundamental principles of whatever he was working on and educating those around him about his liquid crystal theories. Tom thought Jim simplified complex theories involving optics or mathematics so that those he was educating could understand them, but later realized that this wasn't the case. Jim was a visual thinker and his explanations were more a window into his mind.

Jim often talked with Tom about the three fundamental forces that determine how a liquid crystal structures itself. The theory that explains the structures and forces behind them is called the elastic continuum theory, and the three forces are the splay, bend and twist elastic constants. Splay is the strongest force, followed by twist and then bend. A consequence of this hierarchy is that a liquid crystal will bend more easily than it will twist and twist more easily than it will splay. A small degree of splay can store as much energy as a greater degree of twist or bend. The structures formed by these forces are dynamic and can be undone and formed again.

Jim explained to Tom that liquid crystal molecules are more attracted to each other along their sides than on their ends. Tom put "along their sides" and "elastic" together in his mind and built a physical model of a nematic liquid crystal using two blue rubber bands, 18 yellow sticks and some thread.

Model of a relaxed or unenergized nematic liquid crystal

The sticks in the model represent molecules. The rubber bands represent the elastic constants that tug at the edges of the molecules and keep them aligned parallel with their neighbors.

The model intrigued Jim and he used it to demonstrate how splay works. He held the two end sticks and tipped them away from each other. The top rubber band stretched, the bottom rubber band relaxed, and the sticks in the middle aligned themselves along a gradual arc from one outside stick to the other.

Model of nematic liquid crystal demonstrating splay

Jim then lifted the model up and rotated the top stick through more than a full turn to illustrate the twist in a nematic. He explained that individual nematic molecules are bent at one end and this causes the twist to occur naturally, without the need of an external force.

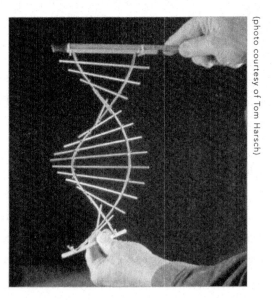

Model of nematic liquid crystal demonstrating twist

ALFRED SAUPE COMES TO KSU

Another scientist who had a deep understanding of the molecular physics of liquid crystals was Alfred Saupe, who Jim had first met at Westinghouse. A well-known liquid crystal scientist, Alfred was fired from his position at the University of Freiberg in West Germany following the death of his former academic advisor and colleague, Wilhelm Maier. It was difficult for even brilliant theoretical liquid crystallographers to find employment. Alfred told Jim he wanted to work at LCI, and Jim was eager to have him. Physics professor William Doane, who was collaborating with Jim by investigating liquid crystal materials that Jim had synthesized, backed Alfred's hiring. Alfred started at KSU in late 1968, when he was 43 years old, buying a house for his family just down the road from the Fergasons. He received tenure as a professor of physics a year later.

Alfred's arrival began an intellectually exciting year for Jim. Alfred was a genius and the most educated physicist and knowledgeable liquid crystal researcher Jim had met. After Max Garbuny, he had the biggest influence on Jim. Like Jim, he did not believe in the existence of swarms to explain liquid crystal behavior. His reputation continued to grow throughout the years and he became very influential in the liquid crystal community, receiving many awards and becoming known as the "grandfather of liquid crystals."

Alfred was drafted into the German army in 1943 when he was in his last year of high school, but was captured in the Netherlands in 1945 and spent three years as a prisoner of war in England. He spoke English softly but firmly, only occasionally stopping to search for the right word. He and Jim had lengthy, energetic discussions on focal-conic textures and the three elastic constants. Jim also observed the optical phenomena in his Leitz polarizing microscope and performed experiments with field effect devices. Alfred performed a large number of experiments he could have written about, but he wasn't interested in reporting what he considered minor discoveries and he never wrote a book—he preferred to talk.

To take a break from the labs, Alfred, Jim, Ted Taylor and Tom Harsch would stroll up Lincoln Street to the Perkins Pancake House for a mid-morning coffee break, often bringing along one or two of the graduate students who were working at LCI. They would linger over their coffee, enjoying the pleasant atmosphere and discussing liquid crystals, the day's work, and upcoming plans. Alfred was well connected in the European liquid crystal community and reported on the latest news from that front. He wasn't shy about designating research he considered of poor quality with the ultimate condemnation, calling it "silly." Sometimes the group lunched on homemade turtle soup at Eddie's, a local bar. They also frequented the Camelot for their mini-surf and turf special: a small lobster tail and filet mignon for only $1.95.

JIM'S EXTENSIVE PRESENTATIONS

Jim made great efforts then and throughout his career to inform the public about developments in liquid crystal science. He made over 50 presentations on basic liquid crystal technology at colleges and companies across Ohio. He also authored an article describing experiments featuring the thermal mapping capabilities of cholesteric liquid crystals aimed at students interested in studying liquid crystals. In each, the thermal visualization device was a Mylar membrane coated with a cholesteric liquid crystal. The experiments proved popular. In part for these educational activities, the Optical Society of America awarded Jim the David Richardson Medal in 2007 for "outstanding contributions to the understanding of the physics and optics of liquid crystals."

8. SUCCESSES AND CHALLENGES AT LCI: 1968

"Although [Jim's] success may possibly be attributed to a certain element of luck, you will agree with me that, in the long run, the really good physicists have all the luck."
—Max Garbuny writing to Dr. Glenn H. Brown

IN 1968, JIM, Sardari Arora and Ted Taylor began investigating the phase transitions of a group of liquid crystal materials. They observed how the molecules reacted when the parameters of variables such as light, temperature, electric charge, chemicals, or substrate material were changed.

These materials were designed to have both a large negative dielectric anisotropy and a high birefringence. The molecular axis of a material with negative dielectric anisotropy aligns perpendicular to an applied electric field. Birefringence is the refracting of light into two separate rays and is a property of crystalline materials such as quartz and LC materials. LCs are considered to have high birefringence because the difference between the two indices of refraction is greater than in materials such as diamonds, glass and water, which have only a single index of refraction.

The materials were created by lateral substitution of chloro and methyl groups into the molecular structure of liquid crystal molecules. During an examination of the phase transition of the materials with regular (not plane polarized) light using a hot stage microscope, Sardari observed an anomaly. The material changed from one phase to another without the corresponding visual effect usually seen in phase transitions.

Jim decided to examine the materials under polarized light. Polarization narrows the parameters of the light so that it reflects off the materials, enabling Jim to observe small areas of birefringent colors rippling like a wave as the materials passed through the phase transition. Under regular light, the phase transitions were occurring, but the colors were not visible. When the sample was slowly heated and cooled over a narrow temperature range around the phase transition temperature, Jim could see a visual effect in areas where the molecular alignment of the LC molecules was changing. He concluded the materials had two liquid crystal phases. Both displayed similar nematic textures, but were separated by a first order (continuous phase) change.

When Jim and Sardari discussed these observations with Alfred Saupe, he referred them to work by Horst Sackmann at the Halle Liquid Crystal Group at Halle University in East Germany and suggested that the lower temperature liquid crystal phase might be a smectic phase with the director tilted at an angle in relation to the substrate.

The concept of a "director" is fundamental in liquid crystal theory and applies to all three phases: nematic, smectic and cholesteric. The term describes the preferred orientation along the long axes of the rod-like liquid crystal molecules in bulk to various macro parameters naturally or artificially imposed. Liquid crystal molecules are constantly moving, oscillating around their centers and rolling over one another.

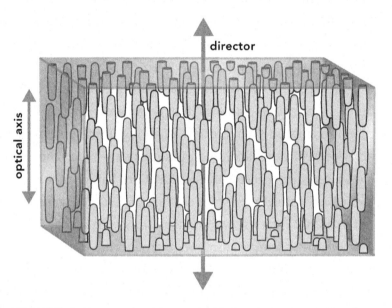

FIGURE 4. Nematic liquid crystal with the abstract concept of the director

Proving that the lower liquid crystal phase might be a tilted smectic phase would be important for two reasons. It would confirm the existence of tilted smectic layers, and it would show that it was possible to impose an orientation of the director across these layers. However, the group was not yet able to confirm that the phase was smectic. They wanted to draw further on Alfred's technical and theoretical knowledge of liquid crystals and asked him to join in the research. Alfred accepted the offer, and made key contributions to building an understanding of the properties of the materials.

Dr. George Gray at Hull University had previously synthesized the same compound Jim and his colleagues were now investigating. But the Gray group had completely missed the nematic–smectic transition, perhaps because they used classical methods to observe phase transitions. Fergason's group was more comprehensive, using both microscopic observation and differential thermal analysis for the inspection. As Tom Harsch told us, "Jim's choice to emphasize visual, or optical methods to study and learn more about liquid crystals made it much easier for him to start thinking about the application of LCs to display devices than it would have been for a person who studied LCs using x-rays or nuclear magnetic resonance, which had no direct visual imagery outputs."

Early in the investigation, the Fergason group discovered that both nematic and smectic liquid crystal structures could be twisted by chiral additives (optically active chemical molecules that are not superimposable on their own image) or by gently rubbing the glass sandwiching the liquid crystal material with cloth. This pioneering research followed earlier research by French scientist Charles Mauguin, who was famous for his study of solid crystals. In 1911, Mauguin conducted experiments with nematic liquid crystals and was the first to observe that contact to a substrate such as glass affected the orientation of the molecules. When nematic liquid crystals were in contact with clean glass, the optic axis of the molecules was perpendicular to the glass, whereas contact with freshly cut mica aligned the molecules along the cleavage plane (planar alignment).

Later, Pierre Chatelain discovered that gently rubbing the glass in one direction also aligned the molecules along the cleavage plane. Mauguin found that applying an external magnetic field perpendicular to the substrates reoriented the long axes of the molecules parallel to the field. Finally, he determined that a layer of liquid crystal was able to affect the state of polarization of light traveling through the layer. In particular, he noted that a properly oriented layer of liquid crystal material could rotate or twist the plane of polarization of linearly polarized light.

Jim and his collaborators explored the use of the new liquid crystal materials. With Ted Taylor, Jim developed a particularly clever means of determining the tilt angle of the liquid crystal molecules in relation to the smectic planes. The diagonals of two ninety degree prisms were unidirectional rubbed. The layer of liquid crystal was placed on the diagonal of one of the prisms and then the diagonals of the two halves joined to form a cube. A polarizing microscope was used to observe the sample.

Being able to measure the tilt angle turned out to be important because the tilt angle influences the performance of liquid crystal displays. At the time, Jim had no thought of making a display device using twisted nematics. Yet he was building the in-depth knowledge of liquid crystal behavior that would eventually enable the twisted nematic and other inventions.

Experiments were conducted on a liquid crystal named DBOAC. It showed a second order, or continuous, smectic A to smectic C transition. In the smectic A phase, the molecules are titled along the director, while in the smectic C phase, the molecules are tilted away from the director. In this compound, the tilt angle of the nematic and smectic A produced by the alignment layer was 90°, and the tilt angle in the smectic C planes went from 0 to 22°. Optically, the difference between the two phases could be observed because the smectic A phase was uniaxial, with the molecules aligned along one axis, and the smectic C phase biaxial. Jim and Ted were able to measure a change in the tilt angle as a function of temperature. In addition, the liquid crystal had a further transition to a smectic B phase. Jim published these discoveries, disseminating knowledge about the new field of liquid crystal science and created a body of work for others to build on.

He also thought that the technologies reported in these three articles might have an application in a dimmable rear view mirror, though not in any display-related applications, which would require inclusion of a polarizer and consequentially result in an unacceptably high loss of display brightness. However, it soon became obvious that the utility of such a configuration would far outweigh the aesthetics, and the technologies discussed in the articles would eventually be embodied in the twisted nematic liquid crystal display (TN-LCD). These articles certainly suggest a major step in the inventive process that eventually resulted in Jim's conception of the TN-LCD.

Up to this point, Jim, Sardari and Ted were researching smectic phases in materials with negative dielectric anisotropy. Nematics that exhibited negative anisotropy at room temperature were common, and Sardari made them often. Liquid crystals are unusual in that they have two values for the dielectric constant: one that lines up along the director and one at right angles to it. Molecules that have a high dielectric constant along the molecule's long axis and a lower constant at right angles to it tend to lie down when a field is applied and are called negative nematic compounds.

Negative nematics were used in dynamic scattering mode (DSM) displays, which were conceived by George H. Heilmeier at RCA in 1968. The liquid crystal material in DSM displays was sandwiched between two glass plates coated with a clean, transparent conductive coating. The molecules aligned perpendicular to this substrate. When an electric field was applied, the orientation of the director changed from perpendicular to parallel to the plane of the glass. By itself, this effect only caused a slight scattering of light, but the electric field caused ions to flow from one electrode to another, and these jostled the liquid crystal molecules, creating twists and distortions in the director. Jostling produced microscopic and chaotic structures. The jostled molecules, which looked like dark, moving worm-like threads under a microscope, scattered the light, caused the molecules to appear darker. This was the dynamic scattering effect. Dynamic scattering devices went from being ordered, with surface alignment determining the orientation of the director throughout the liquid crystal film, to chaotic when the device was switched on.

Neither Jim nor Tom thought dynamic scattering mode displays were much of an invention because the DSM effect was well known. The RCA chemists developed a room temperature nematic with the negative anisotropy that DSM requires. They etched indium-tin oxide (ITO) and/or glass with an evaporated mirror coating, which was not an invention. They worked a surface that the negative LC would orient normally to (the geometric normal), and that simply required a thorough cleaning of the glass (an effect that was well known). Jim and Tom never figured out what RCA "invented."

Jim asked Sardari to synthesize a positive nematic liquid crystal with positive dielectric anisotropy. Making a positive nematic liquid crystal would require reversing the dielectric constants so that the molecules would have a low dielectric constant along the molecule's long axis and a higher constant at right angles to it. Molecules with positive dielectric anisotropy align parallel with, or "stand up," in relation to an electric field. Positive anisotropy nematics were rare, and one that was positive at room temperature had never been made.

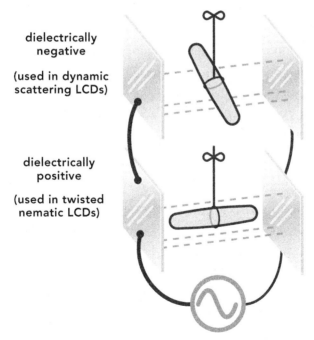

FIGURE 5. Orientation of the dielectrically negative and dielectrically positive liquid crystals in an electric field

Jim and Sardari's positive room temperature nematic opened a door to electro-optic experimentation. It allowed electrical control of the index of refraction, which is the underlying basis of liquid crystal displays.

A REJECTED ARTICLE AND SOUR RELATIONS

Jim conceptualized devices that used the positive material in which the inside surfaces of the cells were "aligned" with the grain stroking the glass in the same direction, as opposed to the to-and-fro rubbing of the glass against a cotton wheel. Tom built these devices, and observations with a polarizing microscope made it clear that the liquid crystal director was aligned with the common axis.

Tom made a device with the positive material and parallel alignment, a variable birefringence cell. It seemed more elegant than previous devices because it was orderly. Without an electric field, virtually every molecule in the entire liquid crystal layer was oriented in the same direction and parallel to the glasses. Then, as an electric field of increasing voltage was slowly applied, the molecules gradually "stood up" until they were perpendicular to the plane of the glasses. The motion of the liquid crystals director could be observed through a polarizing microscope. The men were excited to realize that they had achieved control over the director that, in the dynamic scattering displays, only squirmed and wriggled like fishing worms in a can.

This experimentation was a departure from principles that relied on disorder and a movement toward displays based on ordered effects. It was a "field effect display" (a phrase actually coined later by Jim and his future Ilixco partners), one that was based on optical principles that relied entirely on effects caused by an electrical field. Flowing ions and the current associated with them would be a relic of the past.

The group also researched the use of a layer of liquid crystal contained between two parallel substrates, such as glass substrates. In this configuration, the director was generally found to vary continuously. In practical liquid crystal devices, it was necessary to control this natural variation in the director since device operation required the creation of physically large areas in which the director is uniformly aligned.

Sardari, Jim and Albert submitted an article to the Second International Liquid Crystal Conference held in 1968.[18] They identified for the first time a tilted smectic phase, the use of a surface to align the smectic phase, the use of a cholesteric additive to induce a twisted structure and the effects associated with the application of an electric field, including conditions that induced light scattering in both the nematic and smectic phases.

The articles presented at the conference were to be published in a book edited by Glenn Brown. Glenn sent Jim's article for peer review to Joel Goldmacher at RCA in Princeton, and Goldmacher wrote a letter to Glenn rejecting the article. Articles are commonly rejected because of problems with the content or research, but Goldmacher rejected it because he disliked the title. He made no mention of the article's technical content. As a result, Glenn refused to publish it.

The scattering effect reported in Sardari, Jim and Albert's article was the same as that underway at RCA. Jim speculated that the real reason why Goldmacher rejected the article was to give RCA time to make a public announcement of its dynamic scattering (DS) display. The article's publication would not only have usurped RCA's announcement of their development of DS displays, it would have devalued the impact and public relations value of the announcement. It's also possible that Goldmacher told other RCA employees about the content of the article.

Alfred Saupe, defending Jim, confronted Glenn about rejecting the article on the triviality of a supposedly poor title. Ultimately, Glenn allowed the article to be published in the conference proceedings, but relations between Jim and Glenn soured further. Sardari made an oral presentation at the conference on the group's findings. Afterwards, a scientist named Wolfgang Helfrich, aware of both the article and presentation, wrote an article on chiral additives reporting similar results. The article was submitted to *Physical Review Letters* and sent to Alfred Saupe to be refereed, but it was rejected and never published. Helfrich would end up playing a prominent role in the controversy over invention of the TN-LCD, claiming the invention for himself.

RCA's dynamic scattering display was the first liquid crystal-based display device that RCA intended to market, although this marketing never occurred. In the fall of 1968, the *New York Times* wrote an article about liquid crystal-related developments at RCA. The reporter called Jim for comments. The one comment Jim gave was that RCA's announcement was important because it supported the development of a room temperature nematic.

LCI AND THE VIETNAM WAR

The political climate in the country had been changing dramatically since 1965, when President Lyndon Johnson sent ground troops to Vietnam after passage of the Tonkin Gulf Resolution by Congress. While many in the U.S. supported the war, there was also a growing anti-war movement. Protests were initially strongest in big cities and on college campuses. During the next several years, as Johnson committed more ground troops to Vietnam and increasing numbers of men were drafted by the Selective Service, opposition to the war grew.

College students were exempt from the draft and this led to criticisms that the poor, who couldn't afford to enroll in college, were being disproportionately sent to fight. Martin Luther King, Jr. publicly spoke out against the war in his 1967 speech "Beyond Vietnam—A Time to Break Silence."

In 1968, the peak of America's involvement in the war, protests erupted nationwide. In August of that year, riot police fought thousands of anti-Vietnam war protesters marching outside the Democratic National Convention in Chicago. The institution of a draft lottery in 1969 galvanized protesters on college campuses, particularly campuses visited by military recruiters. Draft resisters burned their draft cards or turned them into Selective Service and protested the ROTC. An estimated 50,000 draft dodgers fled to Canada.

Heated anti-war protests took place at Ohio State University and Miami University, but not at Kent State University. KSU students were by no means "radical," as the results of a 1970 survey of the student body on their opinions about the war in Vietnam and the presence of ROTC on campus revealed. The first organized protest that occurred on campus was a civil rights protest in 1968 by a group called the Black United Students. They demonstrated for the establishment of an office for student affairs which would specifically address black students' concerns. A branch of the anti-war group Students for a Democratic Society (SDS) formed on campus, but had limited success recruiting students. The group remained small in number, comprising at its peak only about one percent of the student body.

Against this backdrop, Jim was preparing a proposal for a Themis grant, sponsored by the Department of Defense, to establish so-called "Centers of Excellence." Most states provided little funding for science research at universities, and Ohio was no exception. The military, however, because of its needs during World War II, was a major provider of support for research. The Department of Defense had a policy to fund general research to sustain a scientific community in the U.S. which Jim teasingly called a "make work program for scientists."

He worked hard on the Themis grant proposal because the funding from it would enable LCI to conduct the basic research needed to fill in many of the blanks in liquid crystal device physics and chemistry. To prepare the best possible proposal, Jim contacted the Engineering Research Laboratory at Fort Belvoir, the funding agency. The person administering the solicitation had been a colleague of Jim's at Westinghouse and told Jim exactly what he needed to write for his submission.

Glenn was still serving as the dean of research in 1968, responsible for reviewing and authorizing any proposal prepared by a faculty member that could bind the university in a contract. Jim sent his completed proposal to Glenn for review. Without telling Jim, Glenn substituted his name for Jim's as the principle investigator and creator of the proposal and then submitted it. Jim was furious when he found out. Glenn justified what he had done by saying that the principle investigator needed to have a PhD to win the contract.

The Themis grant was awarded to LCI in September of 1968, and Jim began working on the project. The Students for a Democratic Society found out about the grant and demonstrated outside the building. They believed that LCI had developed a sensor that was used to hunt down and kill Che Guevara. However, the work done under this grant had no practical application to military detectors and weapons. None of the research was secret, and nobody working at LCI needed security clearance to enter the building. The *Daily Kent Stater* published a letter by a physics professor whose research in theoretical physics at LCI was funded by the Themis grant. He wrote that the Themis-funded research was "basic research on detectors for sensing electromagnetic radiation." Glenn Brown of course knew this, but instead of defending Jim, he told several people that Jim was involved in secret military research.

The rumors persisted. Students for a Democratic Society published a pamphlet listing their demands, among them the abolition of campus ROTC and "elimination of an arcane agency known as the Liquid Crystal Institute."

The Department of Defense declined to renew the Themis grant, as the results were not applicable to any of their programs. However, the Air Force Office of Scientific Research picked up the funding. Here too, Glenn attempted to muscle out Jim as the sole researcher. Glenn mailed back a signed contract for the grant to the Air Force on June 3, 1968, inserting his name as co-principle investigator. In a cover letter, he explained that he was "required to add my name as co-principle investigator since graduate students are involved in the program. Mr. Fergason holds a bachelor's degree which is not acceptable to our university for direction of research being done by graduate students." This was a bald lie, since Jim had long been directing graduate students at LCI.

THE COMPLETE BREAKDOWN IN RELATIONS WITH GLENN

Glenn's meddling pushed Jim to his limit. As congenial and even-keeled as he was, he was no pushover. He formally protested to the assistant dean of research, Daniel Jones, who informed the university president, Robert White, of Jim's protest. Jim's complaints were serious enough for the university to hire Booz Allen Hamilton Inc., a management consulting company, to investigate the management organization and practices of the institute. Jones also contacted the Air Force directly and discovered that they considered Jim the principle investigator for the grant. President White then wrote Glenn, Jim and Jones, "It would seem that an Air Force ruling would sustain Mr. Fergason's contention" that he, and not Glenn, was the principle investigator, adding that, "As principle investigator he should control expenditures and direct technical personnel under that ruling."[19]

After Glenn was told that a Booz Allen Hamilton inquiry would be conducted, he shot off an angry letter to White.

"I have given KSU my best for eight years. I am greatly disturbed that my integrity is being challenged and that you support the accusations of one who has been with us only two years without getting the facts correct. Your action goes contrary to KSU regulations on graduate studies and seems to show lack of courtesy to our research staff who [sic] hold the doctorate. You have embarrassed me by making me subservient to one who holds only the B.S. degree and does not have the professional credentials to carry out what he insists he will do."[20]

Less than a month later, Glenn sent another confidential interdepartmental memo to White. He had apparently contacted the Air Force office and managed to muster support from someone there, because he wrote, "The official contract signed and approved by the granting agency has Mr. Fergason and Dr. Brown as co-principle investigators on the project. I checked with Dr. Elliott on this matter and he assured me that the contract calls for the joint appointment and that Mr. Fergason is not recognized by AFOSR as the only principle investigator. Dr. Elliott said he did not have a copy of the official contract when Mr. Fergason visited with him. It is clear that nothing improper happened on the contract in our relationship with the Washington office as had been stated by Mr. Fergason. The lengthy exposition by Mr. Jones raised 'ghosts,' with no foundation, about our relations with agencies in Washington."[21]

A few paragraphs later, Glenn also wrote. "I have refrained from mentioning this item before, but I feel it has such a bearing on this situation that something needs to be said. Mr. Fergason has an explosive temper and does not make an effort to control it. After working with him, one soon learns that you can't detect when, or over what matter, this temper response will surface." He added, condescendingly, "Mr. Fergason has many fine qualities if he could only learn to control his temper. I have tried to help him and we are making progress but it is slow."

He ended the letter with "Mr. Fergason and I will resolve the problem," but clearly the breach could not be resolved. The two men had different ethics. Glenn wanted to take credit for all the institute's achievements. He had tried to strong-arm Jim for control of Jim's contracts. What's more, Glenn was stuck in the swarm theory rut and his research was at a dead end.

The results of the Booz Hamilton Allen inquiry, if it did occur, are lost to time. However, the university administration tried to put controls in place to retain Jim and to prevent Glenn from interfering with Jim's contracts. A September 10, 1968 letter from Daniel C. Jones to President White detailed the behind-the-scenes negotiations. Jones wrote, "Mr. Fergason has agreed to continue to work within the institute with a division of work. Dr. Brown has also agreed to setting up a section within the institute devoted exclusively to Mr. Fergason's contracts." He added this caveat, "We still have some details to work out but expect these are not insurmountable. One of the stumbling blocks is the ARPA Program. Dr. Brown is making it very obvious that he would still like to be in charge of this program. I think, however, that he is resigned to the fact that Fergason is principle investigator, but expects that on renewal, a year from now, he will again be made principle investigator."

In his reply, White expressed apprehension about the success of these arrangements. As for Jim, he had anticipated having an academic science career, but knew this was impossible now because he wouldn't be able to get another position without a reference from Glenn. He was under contract with LCI and couldn't quit immediately, but he was more than ready to start his own company and perform applied research on his own time. He asked Tom Harsch, Ted Taylor and Sardari Arora if they would consider leaving the university to work with him and they all said yes.

9. THE LAUNCHING OF ILIXCO

> "Ilixco was truly a small business.
> It was too small to have a garage."
> —Jim Fergason, "Inventing the Future" video, 1993

THERE WAS NO policy requiring Jim to inform the university that he intended to start his own company, but he wanted his break to be clean and wrote a letter to Provost Bernie Hall nonetheless. In it he stated that the company's purpose would be to develop direct applications of liquid crystals and to undertake projects that would be "beneficial to the development of liquid crystals and their future use which, by virtue of policy decisions by the Institute, would not be welcome." He wrote that his work for this corporation would not be carried out on university time and would not conflict with "existing or contemplated university projects." He sent a copy of the letter to Glenn.

He also sent a memo to Glenn around the same time about the contract LCI was in the process of completing for the U.S. Time Corporation, maker of Timex watches. Timex had contracted with LCI to make improvements to dynamic scattering mode liquid crystal displays, and Jim and Sardari worked on that contract.

Timex and other watchmakers at the time used Schiff base liquid crystal compounds. These remained stable in a dry environment, but would break down into components when exposed to humid air. The compounds contained ionic contaminants that were necessary to make them scatter, but the contaminants were a result of the synthesis process and their chemical formulas were not known or well controlled. Some of the contaminants caused sluggishness and the display would scatter in a direct current (DC) field and in low frequency alternating current (AC) fields, but not at higher frequencies of 50 Hz. This trait changed over time, as the electrochemical reactions at the electrodes created new unknown ion species. Because of this, a watch display had an operating lifetime of between just one and two years.

Jim and Sardari devised a scheme using molecular sieves to remove all of the ions left over from synthesis. They then added known ionic species, an improvement that allowed higher frequency and more reliable operation over a longer life. However, this did not reduce the electrode deterioration under DC operation. Dynamic scattering displays had a very gradual turn-on (scattering vs. voltage curve). The turn-on is how quickly the display brightens when voltage is applied. Jim created a double layer cell device that turned on more sharply (though this was just an incremental improvement compared to the fast turn-on, at a low voltage, of the twisted nematic (TN) device he would invent two years later).

On April 25, 1969, Jim sent a memo to Glenn and Dr. Bernard Hall titled "Potential Patent." "In accordance with the Kent State University patent policy as of November 21, 1968, paragraphs B6 and B7, I would like to inform you that we have developed what we feel is a potential patent involving materials which form room temperature nematic phases. These materials were developed under our program with the U.S. Time Corporation and, under the terms of that agreement, may be patented by the University." These materials are referring to the dynamic scattering liquid crystal display materials he and Sadari Aurora had worked on, not the twisted nematic.

Jim's statement about the potential patentability of the work he and Sardari had done for Timex was optimistic. The assistant dean of research, Daniel Jones, smelled money and within days fired off a memo to President White stating that the family of materials Jim developed, "should prove to be quite valuable, and, in my opinion, should be patented immediately."

It's obvious from the memo that Jones had been negotiating with Timex behind the scenes, because he also wrote that Timex, "has agreed that they will license the use of these materials from Kent State paying a royalty for the exclusive use for time-keeping purposes when a suitable watch face is profitable. There are many other applications for these types of material, and it would behoove us to react quickly in obtaining a patent." He ended his memo with, "I have advised Mr. Fergason to write a patent discovery statement to Dr. Hall and Dr. Brown."

In the 1960s, less than 100 patents per year were issued to universities. There were ethical concerns about higher institutions making a profit from scientific discovery, but this began to change nationwide beginning with government support of patenting by universities.[22] Before Jim's arrival at KSU, the university had no patent policy. Jim had started his inventing career at Westinghouse and when he came to LCI, he was already an experienced inventor with many patents to his credit. His arrival may have spurred the university to implement a patent policy so they could get reap potential profits from the discoveries at LCI.

THE NEW COMPANY NEEDS A NAME

It is not clear whether Jim wrote a patent discovery statement at the time. He was busy establishing a materials manufacturing company, Tensor Liquids, while continuing to fulfill his contractual obligations at LCI. He spent $5,000 of his own funds to launch this business, in part by drawing on the equity in his house.

Jim hired Daniel Jones to process the paperwork required to set up the new company and Dr. Peter Pick, then a full-time employee of Hoffmann-La Roche, to synthesize liquid crystal materials in his New Jersey garage. The materials could be synthesized in a garage because the liquid crystals were intended for dynamic scattering applications and did not require high purity.

Four months later, Jim found out that Daniel Jones had never followed through with processing the paperwork for Tensor Liquids. Jim still wanted to incorporate the new company, but under a different name. The Fergasons threw a party and invited Tom Harsch, his wife Jacky, Ted Taylor, Sardari Arora and Jim Bell (the banker and venture capitalist from Cleveland who had contacted Jim the year before when he worked at LCI) to think up names over a bottle of scotch.

The group proposed the name International Liquid Crystal Company or "Ilixco," and then someone jokingly proposed naming the company International Crystal Liquid Company or "Ixlic," which made everyone laugh. The four men had already formed a consulting group. With a new name and through the efforts of a new attorney, the partnership was soon formalized. Ted, Tom and Sardari were each granted a 1/9 ownership interest in the company and Jim a 2/3 ownership interest. Ted initially balked at these terms, but Jim explained that he had put up the money to start the company and was the senior researcher, and Ted finally agreed. The partners agreed to pay themselves an hourly rate of $7.50 for work done on behalf of the company, although initially they worked for free in the evenings to get the company up and running.

In considering further steps for making a clean break with the university, Jim decided to document all work he had done at LCI relating to displays. An opportunity to do this arose late in 1969, when an editor at the trade magazine *Electro Technology* called and asked Jim to write an article on liquid crystal displays. Jim and Ted Taylor wrote the text of the article, which was rich in mathematical formula and liquid crystal theory, but the editor of the magazine cut much of this out because *Electro Technology* was a magazine for engineers, not physicists and chemists. Tom gathered information about the characteristics of other display technologies to compare with the latest information about potential liquid crystal display technologies, and photographed some of the experimental setups. "Liquid Crystals and their Applications" was published in January, 1970, in what turned out to be the last issue of this journal.

Before publication of this article, Jim was not well-known in the electronics business world. The article began to change that. Jim got 1,200 reprint requests, an unusually high number. The requests came from medical device manufacturers and photocopier manufacturers, among others.

This landmark article remains an important key to modern display technology. In the epigraph, Jim wrote, "Liquid crystals are providing a radically new way of converting electrical, thermal, and mechanical signals into colored viewing patterns. As low-power devices, they promised alphanumeric and graphic information displays that draw power only when the display is charged." The article provided a roadmap for both the optics and electric field properties of future displays. Jim discussed controlled surface alignment and the RMS voltage response, showing the kind of thorough understanding of liquid crystal optics that only someone with experiential knowledge would have.

This is a comprehensive account of Jim's experiments with a dielectric positive room temperature nematic. Jim's description of it suspended from a thread is wonderful imagery (the word "nematic" originated from the Greek word for thread). Additionally, he said that, "variations of the thickness and layer will not affect the device's uniformity," which was an important trait to recognize. It meant that when the twisted nematic was discovered, it, too, would turn on uniformly, which is particularly important to multiplexing (using one signal source to drive several display segments).

Jim had been considering electric field effects in liquid crystal materials for quite some time. An article he had published in August, 1968[23] included investigation of a helically twisted structure and the ability of such a structure to rotate the plane of polarization of linearly polarized light. This structure was not quite the twisted nematic liquid crystal display, but heading in the right direction. In the *Electro Technology* article, Jim inched even closer to the twisted nematic idea. And yet, it was not an inevitable step that anyone could have taken. It required a big leap of imagination.

INVENTING THE TWISTED NEMATIC LIQUID CRYSTAL SHUTTER

Soon after writing the article, Jim made that leap in thinking when he conceived the twisted nematic liquid crystal shutter. He put together two disparate ideas to do so: an optical waveguide with a new way to rotate polarized light. He realized that a twisted nematic liquid crystal structure could rotate the plane of polarized light in a manner similar to the way that fiber optics bend the direction of light around its twists and turns and keeps it within the fiber.

Jim originally felt that a liquid crystal-based display device should not include polarizers, because too much light would be lost. However, when RCA and other companies had made liquid crystal displays devices that didn't utilize polarizers, such as the guest-host and dynamic scattering displays, they had poor optical properties and were unreliable. In a sense, this left no other possibility than a liquid crystal display device with polarizers.

Jim first explained his new concept at a Christmas party the Fergasons hosted at their home in December of 1969. It took only several minutes for Jim to describe his idea of making a liquid crystal device with a twisted structure to Tom Harsch while they stood in the kitchen. Tom was excited, and thought how wonderful it would be if the idea worked. With typical confidence, Jim was positive that it would.

"That's how conversations were with Jim," Tom said. "He was so convincing that you were swept up by his enthusiasm and believed what he believed. Of course sometimes there were problems with his ideas, or he was just flat out wrong, but an experiment would reveal that soon enough and then Jim modified his original idea or moved on to the next one."

Tom was eager to test Jim's concept by building a test liquid crystal cell, but had to wait several weeks because the company was busy setting up a new office in a small building above the W.W. Reed & Son Real Estate Company on 114 E. Main Street, within walking distance of LCI. Jim had rented half of the top floor, about 1,800 square feet.

Move-in day was January 1, 1970. Jim's wife Dora and Tom's wife Jacky donned gloves and cleaned the new place for the "boys," scrubbing the floors and walls, vacuuming the old carpeting and cleaning the windows. Jim transported the chemicals he had in his garage at home to the new office. "It was a happy day for my wife not to worry about the large containers of raw materials to synthesize liquid crystal," he said. "I also ordered a portable hood from Fisher Scientific. Glenn had made it clear we were not to do applied work at his

Institute. Independent of the limitations, I was in business!" Like the legendary birth of giant Hewlett Packard in a garage, the current multi-billion-dollar international liquid display industry started in this small facility.

The top floor of the building had been laid out for its first tenant, an internist from South Africa who, coincidentally, was the Harsch family's doctor. The second tenant had been a photographer. Two windowless rooms that could be made pitch-black by closing the door were ideal for performing optical measurements and there were sinks with running water for mixing non-toxic chemicals and cleaning ITO glass plates for cells. Dry chemicals used for mixing liquid crystal materials were stored in a large unfinished room at the back of the second floor. The room was also used, with the windows open, to etch electrode patterns onto ITO glasses with concentrated hydrochloric acid and zinc dust. A central office space with a large 12-foot by 6-foot window facing east let in plenty of daylight, which kept everyone cheerful.

The four partners all still had their day jobs at LCI and paid their mortgages and fed their families on this income. They worked at Ilixco in the evenings without compensation. Jim used the $5,000 he had paid in from his own funds to buy electronic and chemical test equipment. He spent $2000 of this amount on a Monocular Nikon polarizing microscope. He always got the highest quality instruments he could afford because he believed that the best instruments yielded the best information. At LCI, he'd had a $12,000 microscope (in 1960s dollars) that he and Ted used to make observations of smectic liquid crystals. Tom and Jim had taken micro-photographs of liquid crystals with the microscope and blown them up to 11-by-14 inches to show off to visitors.

After the move into the new space, Tom constructed the first twisted nematic cell. He described the process:

> I cleaned a pair of 2-inch by 2-inch square glass plates and rubbed the electrode surfaces against a velvet cloth, the backing of which was glued to a board. I then cut two narrow strips from the cellophane wrapper of a cigarette pack to use as spacers and laid them down along the opposite edges of one of the glasses. With a pipette I drew several drops of the positive liquid crystal that Sardari had synthesized from a bottle and deposited these onto the center of the glass. I slowly lowered the second glass, with the electrode side down, onto the first and spread the liquid crystal out until it reached the edges of the cell. Then I clamped the assembly between the rubber-coated jaws of laboratory glass tubing and secured it to an upright stand over a light box.
>
> Next, I arranged a pair of HN-38 polarizers, one above and one below the cell. And, indeed, it behaved just the way that Jim said it would. I had never seen anything like it. There were no colors. The rotation was complete. Every visible wavelength was rotated through 90°. I wanted to see the opposite optical effect and rotated the top polarizer by 90°. The shutter was now open.
>
> I rotated the polarizer back to its original position and the cell was closed again. I used alligator clips to

connect the liquid crystal cell to a Variac, which is a variable auto-transformer that was capable of producing output voltages from zero to 130 volts AC [and] switched it on.

I watched the cell while slowly turning the large black knob to increase the voltage [...] But this time the twisted nematic cell suddenly opened at about 5 volts. This was startling. None of the other liquid crystal devices, including dynamic scattering or DAP (deformation of aligned phase) had behaved this way. I increased the voltage further and it simply stayed open. When I decreased the voltage the optical cell quickly closed. [...]

I walked down the hall to find Jim. I showed him the effect and he was happy and played with it, too. Then we tried to test the cell's response time. Would it be quick? I disconnected the transformer, turned the output up to 10 volts and touched the disconnected lead to the cell. It snapped open fast! However, we couldn't perform a good test of the cell's turn-off time by disconnecting the wires; the liquid crystal cell was a capacitor and without a discharge path, the voltage across the cell decayed slowly, in about a second. This slow turn-off, which occurred because the cell didn't have a discharge path, was another sign the device was a field effect device.

After spending a few minutes savoring his invention, Jim walked briskly to his office. Halfway down the hallway, he hopped up, tapped his heels together and shouted, "Eureka!"

What was it about the collapsing twist that excited him? The ability of nematic liquid crystals to freely rotate is an important concept to understanding how the twist collapses. The mechanism of collapse was one of Jim's key conceptual insights into how a twisted nematic device would behave.

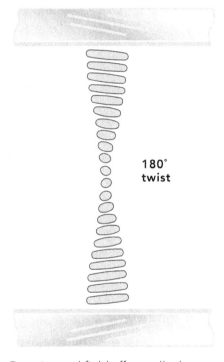

FIGURE 6a. Experimental field effect cells demonstrating twist

In the absence of an electric field, the TN-LCD deformation consists of pure twist. When a field is applied, the first change occurs in the middle of the layer where the director begins to tilt toward normal orientation. This causes a small degree of splay which is projected into the regions above and below the middle and causes bend deformation and a structure similar to the bend-splay balance in Figure 6b.

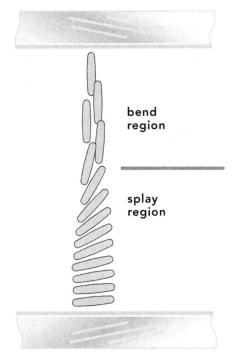

FIGURE 6b. Experimental field effect cells demonstrating splay

As the field increases the center swings though a substantial portion of 90° and reaches a threshold at which the twist cannot be sustained because it is able to release stress by freely rotating around the axis of the director. The twist unwinds and the TN-LCD closes. The twist does not unwind completely unless the voltage is very high. A portion of it remains.

Splay, bend and twist can be thought of as "springs" that attract each molecule to the molecules around it, but are not attached to points on them. The molecules are free to move in three directions, oscillate around their centers and spin. The director(s) that represent the structures formed by splay, bend and twist can also be thought of as having some of the properties of a spring: the structures resist distortion by external forces but removal of these forces cause the structures to "snap back." Like a spring, the forces of these elastic constants store energy.

Figure 6c represents free rotation which means that there is no resistance to rotation around the axis of the director. If the structures represented in figures 6a and 6b were rotated there would be slight resistance, due to the twist elastic constant. After released, the structures would snap back to their original configuration. However, if the structure in Figure 6c was rotated, there would be no resistance and the structures would not snap back.

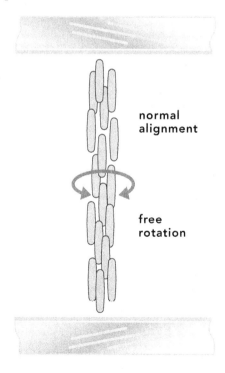

FIGURE 6c. Experimental field effect cells demonstrating free rotation

Jim knew that Charles Mauguin had observed polarization rotation by a twisted nematic cell through a microscope. However, Mauguin observed rotation in a very thin sample of a specific thickness, not a cell that operated over a wide range of thicknesses. He applied a magnetic field to his twisted nematic cell and observed that the field unwrapped the twist. The vacuum tube had not yet been invented. Mauguin therefore did not apply an electric field, or observe the twist unwrapping, i.e., turning on. He made his observations long before anyone thought about making displays.

When Tom turned on the TN for Jim, Jim understood what had happened inside the cell, but he hadn't been sure the twist would actually collapse. That's why he joyfully shouted out "Eureka" when it did.

TIME FOR INVESTORS

With this success, Jim started looking for investors. All good entrepreneurs need the ability to market and sell their ideas, inventions and products, and Jim had this skill in abundance. Naturally friendly and charming, he was great on the phone and easily made contacts with potential customers or business partners. A gifted showman, his demonstrations convinced many customers to buy his devices. He was also able to adapt his subject matter and language to his audience, talking engineering to an engineer, exotic science to a scientist and business topics to management.

Tom Harsch demonstrated the latest TN-LCD device to Jim Bell. Bell was impressed and became Ilixco's first investor, providing $50,000 in 1970. This seed financing helped attract other investors. Bill Osborne Jr., a wealthy Cleveland friend of Bell's who was the chief operating officer of Y&O Coal, invested $250,000. Bell later brought in other investors, including Arora Ventures, which provided $3 million.

Jim Bell served as the chairman of Ilixco's board of directors. Bill Osborne, a Yale graduate with an MBA from Harvard, also served on the board. John Burlingame, a partner at a well-known Ohio law firm, Baker Hostetler, served as secretary and non-voting member. The board met regularly and everyone got along well.

FIRST CUSTOMERS

Ilixco needed business contracts to bring money in. Their most important first customer was Donnelly Mirrors, the supplier of all of the auto mirrors to the Ford Motor Company. At LCI, Jim had established a connection at Donnelly through a colleague, Dr. Walter Lewis "Lem" Hyde, an optical physicist who worked for the J.W. Fecker Division of the American Optical Company in Pittsburgh. As a consultant for this company when he was at LCI, Jim had designed an optical system based on germanium that had properly accounted for the material's high index of refraction. Lem was so impressed by this that he tried to convince Jim to become a lens designer, but he didn't succeed.

The men became friends and Lem later joined the optics faculty at the University of Rochester. Jim got to know a graduate student of Lem's, Dwane Baumgardner, who became the chief technical officer at Donnelly Mirrors and later the president. The $20,000 contract for Donnelly was a study of the feasibility of using a guest-host liquid crystal as an auto-dimming rear view mirror in cars. Ilixco built a prototype of this material, and found that the characteristics of the device would not make it a good auto-dimming rear view mirror.

The company also landed a $40,000 contract with TRW Research Labs in Cleveland to fabricate a machine that used cholesteric liquid crystals to measure the wall thickness of hollow cast turbine blades. This came about because of a project Jim had begun working on at LCI for TRW. They contracted with Jim to design and build a machine to detect structural defects in turbine blade walls. Jim and Tom built an apparatus for testing, but LCI decided to discontinue further work on it so Tom moved it to the Ilixco office for further work.

He and Ted experimented and made measurements to gather data for TRW that showed it was feasible to use cholesterics to measure the wall thickness of hollow cast turbine blades. They delivered test results on six blades to Dick Honeycutt, a TRW physicist. He sectioned and measured them with ball and pin micrometers that confirmed Ilixco's apparatus could respond to changes in temperature in less than 1/1000 second and visualize a 5-6°C range of temperature variation. This in turn could reflect wall thicknesses with an accuracy of two 10/1000 of an inch to a 2-sigma level of repeatability. Honeycutt reported that the statistical variation of Ilixco's measurements was less than the micrometer ones. This new method not only replaced industry standard quality control methods for turbine blades, but improved upon them by providing additional visual information. Innovative aspects of the machine were disclosed in a patent Jim and Tom jointly filed in 1972 and received in 1974.[24]

At some point, Jim asked Tom to make a liquid crystal device using a sample of the positive nematic liquid crystal that Sardari had synthesized. Tom knew what behavior Jim expected from the positive nematic and so did the obvious thing—he made a cell with parallel alignment, placed it between two polarizers that were aligned with the rub direction, applied an electric field, observed the light from a lamp passing through the device and noted that nothing much changed. Then he switched the cell off and serendipitously reoriented the polarizers at 45° to see if something more interesting might happen. Colors appeared that were similar to those Tom had seen when he rotated Mylar film between crossed polarizers, interesting but not unusual.

Tom then applied an electric field and the colors changed in proportion to the voltage applied, from pale pastel colors to red to yellow, to green, to blue, to violet, and finally to dark gray. Turning the voltage down reversed the order. It was a well-controlled experiment with unambiguous results made possible because it was a field effect.

Jim explained that the liquid crystal was acting like a retardation plate, a conventional optical device that splits a beam of polarized white light into two beams and retards one relative to the other. When the two beams exit the plate they combine at a different polarization angle. The effect was the same as Newton's famous refraction and dispersion experiments with a beam of light and a prism, but applied to polarized light to rotate instead of bend it, and disperse the colors around an axis of polarization. The effect caused dispersion, the color constituents of white light spread out into different angles of polarization. This in itself would not appear colorful because our eyes are stimulated equally by all angles of polarization. But a second polarizer revealed the hidden colors. It absorbed polarized light around a certain axis and passed light around an axis orthogonal to the first. The result was that certain portions of the spectrum were absorbed, while others were transmitted.

What made this liquid crystal device fascinating was that the degree of retardation could be changed by changing the voltage applied to the cell. A range of voltages "played" the full range of colors. Jim dubbed the device an (electronically) Variable Retardation Plate (VRP). It was a very pretty effect, particularly because some colors changed quickly in comparison to others. They then had the idea of mounting the cell in a slide projector and driving the VRP with music amplified so that the loudest portions would reach 70 volts. Tom darkened an LCI lab and projected the image against the wall. It was quite dramatic. The device they later called the Videsonic was born. The company self-funded the development of this device, electronically driven by the pulsating beat of music and marketed it to bars in Kent as a "drugless psychedelic experience."

ADVANCING THE UNDERSTANDING OF LC'S

Jim wanted to develop an improved understanding of the fundamental mechanisms underlying the dynamic scattering effect, with the aim of determining if dynamic scattering had commercial potential. He guided Sardari in the synthesis of a series of room temperature nematic liquid crystals capable of dynamic scattering. Ted Taylor and Jim explored a variety of dynamic scattering device configurations, including the addition of various dichroic dyes, utilizing normal alignment, parallel alignment and normal-parallel alignment. All cells were tested by observing the device through a microscope and applying an electric field. None of these variations were found to be as promising as a conventional unaligned dynamic scattering cell.

Jim and Sardari developed an unusual liquid crystal material that had dielectric anisotropy that equaled zero. The alignment of the director of the test device used in the experiments was parallel to the plane of both substrates. When an appropriate electric field was applied to the device, scattering was observed. Interestingly, the scattering effect was observed for only one polarization of light. The results conclusively proved that dynamic scattering was not a field effect. Rather, only electrical current and an alignment layer were required to induce the scattering effect. The results also showed that because current injection was absolutely required, the dynamic scattering effect would ultimately prove unsuitable for use in commercial display devices.

GETTING SHUT OUT OF HIS POSITION AT LCI

By 1970, Glenn Brown's relationship with the chemistry department chairman had become so acrimonious that the university administration decided he could no longer act as the research dean. Likely as a way to keep him in line, they offered him the title of reagent professor of chemistry. He was allowed to keep his position as director of LCI.

Jim's employment contract with LCI was due for renewal by July, 1970. The history of enmity between the two men and the conflict over the Themis grant had made Glenn determined to fire Jim. According to university policy, the decision to fire someone in Jim's position had to be made by a management committee composed of four persons.

Glenn chaired the committee. The other three committee members were chosen by election. Two LCI colleagues were elected to the committee, Adriaan De Vries and Alfred Saupe. For the fourth member of the committee, Glenn only considered candidates from the physics department and not his own chemistry department. He was still lobbying to get his old job back as the chemistry department chairman, but was feuding with half the department and couldn't count on their support. Bill Doane was elected from the physics department.

After deliberations, the employment committee members reached a compromise with Glenn. Jim was at the Ilixco office in late March of 1970 when Dora called to tell him that a letter from the KSU provost had arrived. The letter read: "This letter is to inform you that your appointment in the Liquid Crystal Institute for the year beginning July 1, 1970, and terminating June 30, 1971, will only carry the title of Research Associate. We also wish to note that the Liquid Crystal Institute, upon the recommendation of the Director and the Executive Committee, will not offer you a reappointment after June 30, 1971."

In effect, Jim had been fired but with a one-year notice. He confronted Glenn. Glenn tersely repeated that Jim's appointment would not be renewed. He said that since he was no longer the dean of research, he would be at LCI full time and for this reason no longer needed an associate director. He also said that Jim was heavily involved in applied research and LCI was going to do more basic research. While it was true that Jim had done applied research, such as the breast cancer detection program and the display for Timex, he'd also done pure scientific research, in particular the discovery of a new smectic phase with Alfred and Sardari.

Jim decided to refuse the appointment, but chose to say nothing immediately lest he utter something he might regret. He wryly remarked to Dora, "I published, and I perished." Glenn proceeded to eviscerate Jim's research team by firing two of Jim's grant-funded post docs, Ted Taylor and Sardari Arora. Jim protested to Provost Hall. Hall listened politely, but said that nothing could be done because Ted and Sardari didn't have tenure. Jim's protest was not entirely without effect. An arrangement was made that allowed Sardari to finish work at LCI that was funded by a government contract.

LCI staff had been aware of the animosity between Glenn and Jim, but now the strife was public. A senior university administrator, James McGrath, convened a special committee to investigate it. Bernie Hall interviewed Jim, Glenn, Ted and Tom separately. Tom was asked to describe anything he knew about the problem between his two supervisors. He didn't reply to this uncomfortable question and was baffled at being asked to criticize his supervisors. The inquiry never came to any resolution because of the Kent State shootings on May 4, 1970.

The shootings were the culmination of several days of protest following President Nixon's announcement on April 30, 1970 that he was going to invade Cambodia. Nixon had said in his televised April 20, 1970 speech "Address to the Nation on Progress Toward Peace in Vietnam" that he planned to withdraw more than 150,000 troops over the next year. The invasion of Cambodia just 10 days after this speech angered anti-war protesters.

Demonstrations were held on college campuses on May 1, 1970. A KSU graduate student burned a copy of the U.S. Constitution to symbolize what he felt was the "murder" of the Constitution because of the invasion, and there were other protests later in the day, including one by the Black United Students. The next day, a crowd attempted to set the ROTC building on fire, and someone—it was never determined who—succeeded. In a press conference from the Kent Fire House, Ohio Governor James Rhodes said that at KSU, "we're seeing...probably the most vicious form of campus-oriented violence yet perpetrated by dissident groups and their allies in the state of Ohio." He declared martial law and sent in about 1,000 National Guardsmen.

On May 4, approximately 3,000 people protested on campus against the Vietnam War and the presence of the National Guard. Jim went to the LCI building that morning to meet the west coast editor of *Electronic Design Magazine*, David Kaye, for an interview on liquid crystals. The guardsmen surrounding the building let Jim go inside, and Kaye, who had driven from Pittsburgh to interview Jim, got special clearance to enter. They had just begun the interview when they heard the sound of shooting. Guardsmen entered the building and told the men they had to leave. Kaye completed the interview of Jim at his house, about one mile from the campus.[25] They didn't find out until later that the National Guardsmen had tried to break up the crowd and that a smaller group of 28 guardsmen turned and fired 61 shots at the crowd. Four students were killed and nine others wounded. Tom remembers Jim being sympathetic to the protesters. The piece Kaye wrote after this experience, "Combat Coverage of Liquid Crystals," was published as a sidebar in *Electronic Design* magazine.

Immediately after the shootings, a university employee drove an ambulance with a public address system installed in it around campus, broadcasting an announcement that the university was closed and that all of the students had to leave. The town of Kent remained under martial law. The family saw military helicopters flying overhead. Rumors flew around town—including that two National Guardsmen had been killed, that the police were on the lookout for Weathermen disguised as National Guardsmen, and that the water supply had been spiked with LSD. A curfew was imposed, and, in response to the rumors, road blocks were set up around town to prevent anyone from entering.

The administration canceled the rest of spring quarter classes, shut down the university for six weeks, and did not allow the staff on campus. The Kent State shootings, as they came to be called, put both the university and town in an international spotlight.

With the university closed, Jim had time to recognize how emotionally drained he was from his battles with Glenn. He still had a lot to do to get his new company off and running, but realized he needed a break, and decided to take his family on a three-month trip to Europe.

10. THE PEANUT CAR EUROPEAN SOJOURN: JUNE–AUGUST, 1970

> *"Travel is fatal to prejudice, bigotry, and narrow-mindedness, and many of our people need it sorely on these accounts. Broad, wholesome, charitable views of men and things cannot be acquired by vegetating in one little corner of the earth all one's lifetime."*
> —Mark Twain, *The Innocents Abroad*

WITHIN A VERY short time, Jim put together extensive personal and professional arrangements for the trip, making appointments at companies and universities in Europe to present lectures on liquid crystal-related topics and to provide consulting services for a stipend. He also planned to network and attend the Liquid Crystal Conference in Berlin. The trip helped Jim raise the profile of his future business and make valuable contacts in Europe.

Jim brought Dora, their children Terri, Jeff and John, and his mother-in-law Fern Barlish on the trip. Before leaving,

Dora sewed herself an entire travel wardrobe in beiges and whites, and for her daughter Terri, a blue plaid pantsuit. The Fergasons arranged for Ted Taylor to housesit while they were away. Their car, a big green, air conditioned 1970 Buick, was put at the disposal of the Ilixco employees. Tom and his wife Jacky enjoyed its luxury and air-conditioning for two weeks until they discovered that the exceptionally large gas tank cost a small fortune to fill.

The family started in London, staying in a small hotel, "The toilet paper in the bathroom is something else (non-skid)," Dora wrote in her diary their first night there. "No soap or washcloth, either. A water heater in the bathroom was called a 'geyser.' We had a lousy dinner. We also noticed they have no screens on the windows. We decided it was because the flies all starved to death." The next day, June 10, was Dora and Jim's 14th wedding anniversary. It was also Prince Phillip's birthday. The first stop was Buckingham Palace, where the band played "Happy Birthday," followed by Broadway tunes. It was difficult to find a place to eat because most restaurants didn't let children inside, and when they did find a place, the English food was dismal by American standards.

Several days after arriving in London, Jim picked up the navy blue Volkswagen Squareback he had arranged to buy for the trip. The teeny size of the vehicle—the Fergasons called it the "peanut" car—and of the family, determined the need for John, then the youngest child, to tour the Old World perched atop a pile of laundry in the rear. Driving proved quite an experience. Dora noted in her diary that the roads were very narrow and the English drivers wild. "They're friendly except behind the wheel," she wrote.

Jim gave two lectures at the first stop on the tour, the Imperial College of London. The college's medical school had a long record of studying human body temperature and found Jim's cholesteric work of particular interest. He also lectured at the Chester Beatty Cancer Research Institute. The rest of the family visited the Tower of London and St. Paul's Cathedral.

Jim lectured on the smectic C phase at the University of Southampton, on invitation from chemistry professor Geoffrey Luckhurst. At Hull University in Southampton, Jim met with chemists J.W. Ernsley and George Gray. Gray's 1962 publication, "Molecular Structure and Properties of Liquid Crystals" was the first major publication in English on liquid crystals.

From Southampton the family took a ferry to Normandy and drove to Paris. Jim gave a lecture in Orsay at the University of Paris-Sud. In conjunction with this, he and Dora received a coveted dinner invitation from liquid crystal scientist and future Noble Prize winner Pierre de Gennes. During the day Jim presented the lecture and Dora and the rest of the family went sightseeing in Paris. Dora thought Jim was going to pick her up at the end of the day and drive her to the de Gennes' house, and Jim thought Dora was going to their host's house on her own by train. Jim got lost driving from the university to de Gennes' house and arrived so late that he made it just in time for dessert. He called Dora at the hotel they were staying at to ask her where she was, and found out she'd been waiting for him to pick her up. To make up for stranding her, Jim took the family to a posh restaurant several days later.

On June 23, the Fergasons traveled from Paris to Montpellier in the south of France, where Jim visited Pierre Chatelain's liquid crystal group at the University of Montpellier's Department of Physics of Crystals, which had a history of research in liquid crystals. The visit was stimulating but limited by the lack of a common language. Afterward the family took advantage of the perfect weather and swam in the blue waters of the Mediterranean Sea. The children didn't want to leave, but Jim had a consulting obligation in Grenoble at the laboratories of Thompson CSF.

Meanwhile, the company sent Dora, Fern and the kids to the Alps in a Citroen with a French driver. As they ascended a hair-raising drive up the mountain, the driver tore around the many blind curves. John made a strange sound in the back seat and Dora was afraid to turn around and look at him. When she finally did, she saw that that he had turned green with car sickness. Dora made the driver stop so she could move him to the front seat with her. The driver raced back to Paris at 160 kilometers an hour, or just under 100 mph. Twice he approached an intersection when other drivers were inching their cars forward, but they backed up when they saw the "blood in his eye," as Dora put it.

The Fergasons drove the peanut car northwards from Paris up to Luxemburg and Belgium and entered the Netherlands. Jim stopped at Philips in Amsterdam to visit their newly acquired operation for producing purified cholesterol from wool grease, also known as lanolin. He gave a talk at their research laboratory in Eindhoven, which had generously supplied cholesterol materials for his research, and then gave a seminar before the Amsterdam Cancer Society. In his spare time, he visited the Rijksmuseum with Dora to see the Rembrandts and van Goghs.

On July 4 1970, the family left the Netherlands, driving through Germany to Denmark and then on to Copenhagen. For the next two and a half weeks, Jim took a break from business to sightsee and spend more time with his family. They toured Kronborg Castle (the famed "Hamlet's" castle), saw the Little Mermaid statue, and took the kids to the second oldest amusement park in the world, Tivoli Gardens. They then drove to the east coast to swim in the Baltic Sea.

From there they went to Oslo and its sparsely populated northern region. The family made stops in Narvik and Alta. On July 10, they pointed the peanut car towards the North Pole and explored the Norwegian coast above the Arctic Circle. Everyone greatly enjoyed this part of the trip. It was a high northern latitude where the sun never sets in summer. Traveling up a fjord, the children counted over 100 waterfalls in one day. The family visited a Lapp village and passed by Lapp camps.

Jim then needed to resume the business part of the trip so the family drove south down through Finland and crossed into Sweden. It was still backwoods country and they spotted reindeer walking along the highway.

On July 23 they visited Hans Tillander, a surgeon and a consultant for AstroMedi Tech in Sweden, with whom Jim discussed the uses of liquid crystals in medicine. Dr. Tillander, his wife and their five children were gracious hosts. They had an Airedale dog that captivated Jim with the tricks it could perform.

Four days later, the family headed back to Stockholm, where Jim spent a month with Dr. Stig Friberg at the Technical High School. He had arranged for the Fergasons to stay in a roomy apartment with a large kitchen in a building for visiting professors. Dora cleaned the apartment and set it up so the family would be comfortable for the month. She tried to cook local foods. Everything was very expensive and came in small packages with Swedish instructions on them.

LEARNING NEW TECHNIQUES IN EUROPE

Jim learned a useful technique during the month he spent with Dr. Friberg. Friberg's lab was researching a technique in which a small sample of a surfactant was placed between a microscope slide and a cover slip, and then a drop of water was introduced. The effect of the interaction of water and surfactant was observed using a polarizing microscope.

Jim later adapted this technique at Ilixco to determine the proportions in a blend of two liquid crystals that form a eutectic. He placed a small sample of the first solid liquid crystal at one end of a glass slide and a small sample of the second solid liquid crystal at the other end, then placed a cover glass on top. He heated the sample until both liquid crystals melted. The cover glass settled, causing the liquid crystals to flow. The spreading droplets eventually met at the center of the slide and mixed together. The line of contact represented the position where the proportions of the two materials were equal. As one moved physically away from the line of contact toward the first droplet, the proportion of the second material decreased linearly going to zero at the location of the first droplet. The opposite was true moving in the direction toward the second droplet. The slide was then very slowly cooled, with care taken to assure that the cooling was uniform.

Eventually, most of the material on the slide solidified. The clearing temperature at the line between the two compounds was measured. At Westinghouse, Jim discovered that the clearing points of liquid crystal blends vary linearly with the mole percent of the two constituent compounds. By using this information, and the molecular weight of the liquid crystals, he could calculate the composition of the eutectic. The slide could also give an indication of stability by finding the temperature at which the LCs crystallized completely.

The family left Sweden and drove to Amsterdam, where Jim presented a paper on his work using liquid crystals for thermography at the International Medical Thermography Conference. He then drove into West Berlin from East Germany through Checkpoint Charlie at the Berlin Wall to attend the Third International Liquid Crystal Conference, which was held from August 24 through 28. Earlier in the year, Jim had participated in making an episode of a television show on liquid crystals. The episode, produced for a Cal Tech TV series called "Interface," featured Jim at LCI, and also included an interview with George Heilmeier of RCA. The show aired in 1970 on over 100 stations. Jim had brought a copy of the 16mm film of the show with him to the conference. Rolf Hosemann, who was chairman of the organizing committee and of the conference and worked at the Fritz Haber Institute of the Max Planck Society, asked to borrow the film. He returned it to Jim later without telling him what he did with it. Jim later found out that Hosemann and Glenn Brown, who was in Berlin for the conference, took the tape to a German TV station, which aired the episode.

On August 26, the family visited West Berlin. At two places they climbed up and looked over the Berlin Wall.

They stopped at a place where the wall consisted of a side of a building, with the windows bricked up. There was barbed wire and embedded glass on the side of the building which formed the wall, and they could overlook Checkpoint Charlie. A later bus tour the family took of East Berlin also made quite an impression. The bus had to drive through a maze to get through the gate between West and East Berlin. Nobody could storm the gate because it was mined. The ravages of World War II were still clearly visible in East Berlin.

"We saw large hills around the edge of the city, and these hills were made of rubble from the bombed buildings and houses in the war. They are landscaped now with grass, but it is really a sight to make you stop and think," Dora wrote in her journal. They visited a park where 150,000 German soldiers and 120,000 Russians lost their lives in the battle of Berlin. Dora and her mother were greatly affected by Kathe Kollwitz's sculpture in the park, "Mother with her Dead Son," a mammoth bronze figure of the mother sitting bowed over her dead son who slumps between her legs. His face, an agonized expression on it, is upraised as if looking beseechingly at his mother as she cradles him, weeping, against her chest.

After the liquid crystal conference, Jim visited the Siemens Research Labs and his old friend Dr. Wilhelm Stürmer in Erlangen. Stürmer house was in a charming town surrounded by the Black Forest. There were microscopes in his living room. He served the family coffee and cake and then took them to dinner at the local castle, where they had sausage, sauerkraut and beer. Stürmer gave Jim a tour of the extensive Siemens facilities and arranged for three lectures. They also visited Nuremberg Castle and old town. On September 1,

Terri's birthday, they went to Switzerland, where they met with chemist Sidney Schaeren from Hoffmann-La Roche in Basel, who gave the kids huge bars of Swiss chocolate. They walked up and down the busy, narrow streets of Basel's shopping district. From Basel they drove through Fribourg to Darmstadt, where Jim met with someone at Merck. On September 5, they flew from the Frankfurt airport to London and several days later, back to the U.S.

Jim and Dora joked that their European adventure had turned Jim into a "migrant worker," though a well-paid one. Jim earned enough in stipends from lectures and consulting that he came back with more money than he had left with. The profit side of the ledger included the little blue Volkswagen Squareback that Jim had squired the family in for 9400 miles around Europe. The family had grown fond of it and shipped it across the ocean.

Jim and Dora, 1973

(photo courtesy of Tom Harsch)

Jim had not forgotten the wonderful Airedale Dr. Tillander owned, and talked about the dog after their return. Separately Jim and Dora decided to get the family a white fox terrier puppy for Christmas. As the holiday approached, Jim told Dora sadly that "he hadn't found one." Dora kept her secret that she had found one and the children and Jim had an early Christmas surprise. They named the dog Asta after the dog in the Thin Man movies.

BACK AT LCI

Once back in Ohio, Jim had to go to LCI, since he was still technically employed there. The Kent State University campus had now reopened after the tragic Kent State shootings, and Jim was able to enter the LCI building. He headed for his office, but felt disoriented when he couldn't find it. While he was in Europe, Glenn had made Jim's demotion to a research associate plain by moving Jim's office to a smaller one on the third floor. He had also taken away Jim's typewriter and Dictaphone and fired his secretary.

With Dora's support, Jim resigned from LCI. He retained an affiliation with Kent State University's biology department as an advisor for the breast cancer and other studies involving liquid crystals, all of which were funded through grants he obtained. For this continuing affiliation, he received a grand "salary" of one dollar a year. This allowed him to retain the University's health insurance, which was better than Ilixco's, for his family.

Jim left LCI for good in September, 1970. He was just 36 years old, and excited about the future of his new inventions and the freedom he had to explore it in his own company.

11. ILIXCO'S DEVELOPMENT OF THE 1ST TWISTED NEMATIC PROTOTYPES

*"In 1970, nobody—including us—
knew that the twisted nematic was going
to be the next greatest display device."*
—Tom Harsch

WHILE JIM WAS in Europe, Tom used the equipment Jim had purchased to explore the electrical and optical characteristics of the new TN effect. Using the microscope, the photomultiplier tube and special drive circuitry, he was able to measure optical density. He also plotted the turn-on curve and relaxation times under a variety of conditions. He reported his findings to Jim, who had called every Friday from Europe to get the latest news.

When Jim returned to the U.S., these working TN devices enabled him to observe the twisted nematic's secondary optical traits, such as how it behaved off-axis. Having actual working devices confirmed his theories about how TN cells

Jim Fergason at Ilixco, 1974

would operate. Nonetheless, when the partners speculated about what they could do with the TN effect, the possibility that it might become a viable display technology was still remote. All they had at the time was 2 inch by 2 inch TN cells held together with clamps or glued together with epoxy. The company certainly wouldn't be able to sell those.

The cells were drab and dark, too. Tom had used dark laboratory grade type polarizers, which had low transmission, in order to achieve excellent extinction, i.e., the blocking of light when a pair of linear polarizers are crossed. A pair of these used in a transmissive display arrangement passed only 21 percent of the light. Used in a reflective arrangement, the transmission was even worse, less than 5 percent, and the cells could only be seen with a bright light behind them.

Tom, Jim and Ted observed two other problems with the devices. The first was that the effect could be seen at just one angle of view. When they turned the display to see if they could view it from another angle, the display faded out and they couldn't see the digits anymore.

The second problem was something Jim termed "optical bounce." A voltage was applied to turn on the cell, but when the voltage was turned off, the cell turned on again for almost a second. This "bounce" was a result of the tension in the TN one-quarter twist, which is a spring. The viscosity of the LC did not absorb the energy in the spring quickly enough. The TN devices bounced in tests for speed of response. A cell could be turned on (opened) more quickly by applying a higher voltage, but the higher the voltage, the more extended the bounce and the more noticeable to the naked eye.

Jim's analysis was that the bounce occurred because the twist was relaxing and overshooting like wheels on a car with worn shock absorbers. The wheels oscillate or bounce when a car has bad shocks. If a shock absorber is matched to the suspension system—the strength of its springs, and the mass of the wheel and suspension parts—the system will absorb shock going over a bump and transfer only a moderate amount of energy to the chassis. The relaxation of the TN twist "bounced" a change in the configuration of the nematic liquid crystal. In the TN-LCD, the twist elastic constant is equivalent to the spring in a car's suspension system and the viscosity of the liquid crystal (the inherent value of which is particular to the molecular design) is the shock absorber.

Prototypes of the early TN cells actually used cut-up cellophane cigarette wrappers as spacers between the two pieces of glass. Even that was too thick. Jim thought that reducing the thickness of the cells by half would tighten the twist and eliminate the bounce—and he began thinking of ways to do this.

Jim knew that the high tilt angle of the director at the cell surface allowed the twisted nematic 'spring' to be looser than it would be if the tilt angle were low. A lower angle would tighten the spring and shorten the relaxation time. These two factors would optimize damping of the system and eliminate the bounce. Damping reduces oscillations in mechanical systems. For instance, when you strike a bell, it rings for a long time. If you touch the bell lightly the ringing subsides, or is damped. If you put your whole hand on the bell it won't ring at all and you have over-damped the system. Jim's intuition about the effect of the tilt-angle was correct, but at that time they didn't know how to lower it.

THE FIRST TN PATENT

The twisted nematic device fascinated Jim, Ted, Sardari and Tom and when they worked on it, they didn't think about who invented what. Instead they felt the device had a life of its own. They all knew that Jim had the idea for it and that most of what they worked on originated with him, but they didn't think of the device as Jim's because he didn't encourage that kind of thinking. It was fun to explore the device and discover its capabilities.

But after observing the first TN devices, Jim knew that he had discovered something genuinely new and scientifically significant. It was time to assign authorship. He and Tom sat down in the 141 East Main Street office to discuss what they had that was patentable and how to write the disclosures. Jim felt that Tom should be the sole inventor of the Videsonic because he had discovered it all by himself. He didn't like the practice at Westinghouse where some of the managers habitually added their names to the patent disclosures by their scientists, even though the managers had not worked on or contributed to the invention. He also felt that if patents with "extra" names were challenged in court, testimony would be a problem because opposing attorneys could examine people who would not be able to explain their contribution, how the invention worked, and how it was developed.

However, the idea for the TN invention was Jim's so Jim was listed as the sole inventor. Jim sent a disclosure documenting his invention to Ilixco's patent attorney, Tom Murray, in September, 1970. Jim had met Murray when Murray worked as an outside counsel at Westinghouse. Murray's partner John Linkhauer, a chemical engineer and expert on chemistry-related patents, undertook the actual patent work. He promised to write and submit the U.S. TN-LCD patent application by October. Jim also applied for patents for the turbine blade testing technique, with him and Tom listed as co-inventors.

Jim then turned his attention back to how the company might use the twisted nematic device. This was 1970, when digital integrated circuits were popular. Tom was interested in logic devices and proposed that TN cells might be used in series to perform 'optical' logic functions. They even created a term, Optilogic, to describe these devices and used it on a data sheet. The Optilogic idea was impractical because electronic circuits were much faster and easier to use, but Ilixco spent several months experimenting with making Optilogic devices, and learned more about the TN effect from them. Jim was willing to explore new ideas, even if seemingly impractical, just to increase his knowledge of fundamental phenomena.

In October, Jim learned that John Linkhauer had not filed the TN-LCD patent disclosure. Linkhauer promised to submit it soon. But in December, 1970, Jim found out that Linkhauer still hadn't filed the disclosure. From then on, Murray did Jim's patent work himself. He filed Jim's U.S. application on February 9, 1971.

THE BEGINNING OF THE PATENT BATTLE

The delay was costly for Jim. On December 4, 1970, Swiss-based Hoffmann-La Roche (hereafter referred to as Roche) filed an application with the Swiss patent office claiming invention of the twisted nematic liquid crystal principle. Switzerland had a *first-to-file* patent law and awarded patents to whoever applied for them first, while U.S. patent law at the time was a *first-to-invent* law. U.S. patents were awarded to inventors who could provide supporting evidence such as laboratory notebooks, prototypes, and witness of reduction to practice that they had conceived the invention first. Jim had all of this evidence in the patent disclosure. If the application had been filed in October 1970, it would have created a written record that Jim invented the twisted nematic and there would have been no dispute over the inventorship.

Roche listed two scientists as the inventors of the TN-LCD, Wolfgang Helfrich and Martin Schadt. Helfrich worked for RCA Laboratories in the late 1960s in George Heilmeier's liquid crystal group, which conducted theoretical investigations of dynamic scattering. Helfrich claimed that he had conceived the principles of operation of the twisted nematic in 1969 and explained the principles to Heilmeier, who showed little interest in the device because it required the use of two polarizers to absorb a large amount of light.

Liquid crystal work was starting to wind down at RCA in the late 1960s in part because RCA's executives doubted the commercial potential of the technology. Heilmeier left RCA in 1970, and later directed the Advanced Research Projects Agency (ARPA), the forerunner of today's Defense Advanced Research Projects Agency (DARPA). As director of DARPA, Heilmeier declined to fund liquid crystal-related display research. Liquid crystal scientist Frederic Kahn said that "(Heilmeier) did more to stimulate liquid crystal work than anyone I can think of, and he did more to retard liquid crystal work in the United States than anyone I can think of—he has both those distinctions."[26]

Helfrich also left RCA in 1970, accepting a position with the new liquid crystal research division at Hoffmann-La Roche's headquarters in Basel. He was teamed with physicist Martin Schadt, and the two worked on liquid crystal display development.

In the fall of 1970, LCI sent Alfred Saupe to Brown Boveri & Cie (BBC), a Swiss electrical engineering company, which was collaborating with Schadt to develop liquid crystal devices. The work was based on the room-temperature liquid crystal material developed by Joseph Castellano at RCA. Alfred had kept in close touch with Jim and mentioned to them that Jim was working on a twisted nematic liquid crystal display device. (The opposite is claimed in some histories of the invention—that the TN effect was demonstrated to Alfred Saupe when he visited Europe and that he brought the idea back to Jim Fergason.) The BBC researchers told Schadt this, prompting Roche to file their patent application for the TN-LCD with the Swiss office on December 4th.

Four days later, Schadt and Helfrich submitted a scientific article, "Voltage-dependent optical activity of a twisted nematic liquid crystal" to *Applied Physics Letters* reporting their development of the TN-LCD. However, it is known that Helfrich had received a copy of the January, 1970 issue of *Electro Technology* with Jim's article in it 11 months earlier, but did not reference it in his own article.

It's possible that the liquid crystal expert who reviewed Schadt and Helfrich's article for *Applied Physics Letters* did not know about Jim's *Electro Technology* article. The Schadt-Helfrich article would have seemed to present new information. The expert approved it for publication and it appeared in the February 15, 1971 issue of the journal.[27]

Jim Fergason felt that despite Schadt's being listed as a co-inventor with Helfrich, Schadt's involvement in the technological development of the TN-LCD was minimal. Jim speculated that Roche wanted to bolster the impression that the device was developed at their facility rather than being brought to it by Helfrich from RCA. He thought they accomplished this by teaming Helfrich with Schadt, who was an established Roche scientist. Schadt participated extensively in directing his employer's TN-LCD patent-related activities. When Roche suspended liquid crystal research for one year in 1971 because, according to Schadt's account, they were skeptical of the new technology, Schadt remained with the company, but Helfrich left.

NO PROOF THAT EITHER HELFRICH OR SCHADT INVENTED THE TN-LCD

Martin Schadt said that Helfrich had brought the idea with him from RCA, and provides evidence that Helfrich understood the configuration and principles of operation of the TN-LCD during the time frame claimed by Roche. However, Helfrich could have developed his understanding of the device by reading Jim's *Electro Technology* article. One of Helfrich's RCA colleagues reported that Helfrich read the article and said to him that, "Fergason did not know what he had."

There are also no written records that Helfrich conceived of the TN-LCD while he was working at RCA. There is only hearsay, but Heilmeier did not recall Helfrich explaining the device to him. Joseph Castellano, who shared an office with Helfrich at RCA, thought that Helfrich had conceived the TN-LCD, but noted that "there are no written documents or notebook entries to confirm the actual demonstration of the effect and its reduction to practice."[28]

In addition, if Helfrich had invented the TN-LCD while at RCA, the ownership of the device would have belonged to RCA. However, RCA never took any legal action against Helfrich or Hoffmann-La Roche to claim the invention. Indeed, RCA had a policy requiring its research scientists to keep detailed laboratory notebooks and in internal correspondence to protect the company against litigation.[29] Hirohisa Kawamoto, an electrical engineer and computer scientist who worked for RCA Laboratories from 1970 to 1980, published an article on the history of liquid crystal displays in 2002. Kawamoto wrote that he asked Alfred Ipri, and later Alexander Magoun, the curator of the David Sarnoff Research Center (formerly RCA Laboratories) to search the David Sarnoff Library for writings by Helfrich. They found 15 reports, but none contained any reference to the twisted nematic and there were no laboratory notebooks by Helfrich in the library. When Kawamoto met Helfrich at Freie University in 1993 and asked him about TN-LCD documentation at RCA, Helfrich said that he had never used a laboratory notebook because his job at RCA related to activities that helped advance understanding of theoretical physics and not to invention.

A science historian who conducted laboratory notebook analysis for his dissertation on RCA and scanned all of the Sarnoff Library holdings before the library closed in 2009, also found no notebooks by Helfrich among the large collection of notebooks by other LCD research staff. If Helfrich had the idea for the twisted nematic, he did not try to convince RCA to make a test cell and prove the feasibility of the idea, which would not have been difficult to do. Martin Schadt writes that he made the first TN-LCD prototype at Roche in 1971, which itself seems strange if they patented it in 1970.[30] Ilixco, on the other hand, made four prototypes over a period of about one year, from 1970–1971. These were the first TN-LCDs in the world.

GOING INTO PRODUCTION

Ilixco first built a 7-digit and a 5 by 7 dot matrix. For each display, a draftsman made black inked drawings of the electrode patterns. The drawings were reduced in size and photographed to produce a film negative. This film negative was used to expose a photosensitive acid-resistant film and the ITO coating was etched on the glass. This coating was a hard, virtually transparent, electrically conductive coating that made excellent electrodes. ITO electrodes enabled Tom to apply high voltage electric fields across thin LC layers while the transparency allowed Jim to observe the liquid crystal behavior. Patterning enabled the application of electric fields to selected areas. Tom cut little "gaskets" from thin, delicate Mylar sheets with a razor blade. He held together the two glasses for these displays by placing paper clamps around the perimeter. Then he trimmed the excess Mylar off, wiped away the excess liquid crystal and applied epoxy around the edges. This made reasonably reliable operating displays, but they couldn't be mass produced because the process was too slow for production and was only effective at keeping out contaminants for several months.

The third prototype TN display was an 8-digit calculator display fitted into a Nixie tube. One of the most popular digital display types at that time, Nixie tubes were vacuum tubes filled with neon gas. The Burroughs Corporation made Nixie tube calculators, and Ilixco naively assumed they would recognize the TN's superiority. To make the TN calculator Jim bought a $400 (in 1971 dollars) Burroughs calculator and Tom took out the Nixie tubes to see how the tubes were driven and multiplexed. He then fabricated a single TN-LCD that matched the Nixie tubes in number and size. Jim took this to Detroit and showed it to Burroughs scientists and executives. They made an offer of $50,000 for the patent and technology, a lowball figure which was easy for Jim to refuse.

As the TN-LCD neared production readiness in 1971, Jim again contacted Donnelly Mirror and negotiated a contract with them to develop and deliver a 2-digit LCD. Ilixco shipped Donnelly a sample of this display, the first TN display to be sold anywhere in the world.

Donnelly decided not to proceed with TN-LCD development because they needed to reduce expenses. In retrospect, this was a giant lost opportunity, because as part of their contract with Ilixco, Donnelly would have been entitled to what became extremely valuable rights related to the invention. Donnelly's decision forcefully brought home to Jim the differences between working in a university, or in the large Westinghouse research lab versus running his own small business. At LCI, Jim could pursue lines of inquiry just because they intrigued him, but now he needed to think about how to generate revenue. For a creative person, this was a restrictive but necessary lesson.

Ilixco's first public debut of its displays was at the Cleveland Electronics Show in the spring of 1971. Ted Taylor presented a paper on the development of the TN-LCD, for which he was awarded best paper in the show. Ilixco exhibited the 2-digit display, the functioning 5 by 7 TN-LCD dot matrix, and the calculator. Viewers saw images of the liquid crystal display projected onto a diffusing glass plate. A polarizer was mounted behind the diffuser, sandwiched against a TN-LC cell with patterned electrodes and a second HN-38 polarizer. These were the first TN digital displays with etched electrodes. They included appropriate drive circuitry, i.e., the 2 digit and 5 x 7 digit displays switched digits and the calculator calculated. Each display was mounted in a metal cabinet with drive electronics, backlit by a strong green 4-watt fluorescent tube energized by a transformer. The fluorescent tube was needed for light, because of the dark, laboratory grade HN-38 polarizers Tom had used in the display. This system could not be utilized in a panel meter or calculator, let alone a watch.

Although Ilixco had rejected and moved on from dynamic scattering technology, they included a tiny dynamic scattering display in their booth to demonstrate the resolution potential of liquid crystal displays. It had static, 7 segment digits a little over 1 millimeter tall. Visitors to the booth viewed the display through a stereo microscope. The Videsonic projector threw bright, undulating patches of deep rose, yellows, greens and blues onto a screen, with the hues changing in time to input from a radio, stereo or microphone.

An electronics parts distributor called Compar had a booth across the aisle. Norm Case, their representative, was fascinated by the TN displays, which were 20 times taller than the LEDs he was showing. Because of his enthusiasm, Ilixco later engaged Compar as a distributor of TN-LCDs. Compar purchased and kept standard TN-LCD displays in inventory. To fill a small order, they supplied the product directly to the customer. They forwarded large orders to Ilixco, who supplied the order and gave Compar a commission.

After the Cleveland Electronics show, Ted Taylor announced that he was leaving the company. The first few years had been entrepreneurial and creative, but the company was now entering an intense manufacturing phase, which wouldn't be much of a challenge for a physicist. Jim, Sardari and Tom bought out most of Ted's stock. To everyone's surprise, he emigrated to

Brazil to join the physics faculty at the Universidade Federal De Santa Catarina. He married a Brazilian woman and started a family.

To further publicize their displays, Jim and Tom attended the Society for Information Display show in Philadelphia from May 4th to 7th, 1971. They showed direct drive and multiplexed TN-LCD displays in their booth. This was, at the time, the first application of multiplexing to a TN-LCD. The purpose of a multiplex drive was in part to minimize the number of drive circuits as well as the number of contacts required at the edges of the display.

One of the displays contained a new, prototype high transmission polarizer. Ilixco had told the scientists at Polaroid that they needed lighter polarizers than the HN-36 and HN-38 polarizers made for sunglasses and scientific instruments. In response, Polaroid developed HN-42 and HN-45 polarizers, the first in a long line of polarizer products designed specifically to meet the requirements of TN-LCD applications.

Fred Kahn, then with Bell Telephone Laboratories, attended the conference. Fred had first heard of liquid crystals materials three years prior when he visited the Xerox Webster Research Laboratory and saw a demonstration by researchers who were investigating the uses of cholesteric liquid crystal materials in Xerox copiers. He was fascinated, and began to read up on liquid crystals, starting with papers Jim and his co-workers at Westinghouse and LCI had written. Fred became a well-known liquid crystal scientist and inventor. He told us, "Jim's work at Westinghouse and the LCI was the foundation stone and inspiration for my own work." The conference was the first time the two met in person. Jim invited Fred up to his hotel room for a personal demonstration of displays. These were the first twisted nematic displays Fred had seen, and he was impressed with the good image quality.

"Jim's TN-LCD demo was a watershed moment for me," Kahn said to us. "Schadt and Helfrich had written a technical paper. But for me, seeing is believing—and I saw a display in Philadelphia that looked good. Therefore, I believed. This increased my confidence in the potential future success of liquid crystals." In a letter published years later in *Physics Today*, Fred wrote about seeing the display at the Philadelphia show, "James Fergason showed me privately a TN-LCD numeric display operating at room temperature and with good viewing angle."

The Philadelphia show debuted Ilixco's first official data sheet. To create the data sheet for the Philadelphia show, Ilixco used a blank sheet from a print shop with simple graphics that included the company's logo. Tom typed in text to describe the TN cell and drew two overlapping curves that illustrated the turn-on-and-off behavior of the TN device.

Finding the right way to communicate science concepts was important to Jim. He was a fan of the 1967 movie *Cool Hand Luke*, and would sometimes quote from it, "What we've got here is a failure to communicate." He didn't want the information on the data sheet to be confusing. After much thought, Ilixco decided to use metaphors and concepts familiar to electrical engineers. They therefore entitled the data sheet "BILEVEL LIQUID CRYSTAL DISPLAYS—Dec 1971" and made several dozen copies of it.

It presented the technical specifications of the two LCD models the company had designed and produced, the BLM-7040, a multi-digit display with 1-centimeter-high digits, and the BLM-7080, a multi-digit display with 2-centimeter-high digits. On the data sheet, Ilixco recommended several DC drive circuits for LCDs to their customers. The data sheet also showed that the company estimated an AC driven display might last for 30,000 hours. This turned out to be a conservative guess. A display Tom made for Tekelec in 1974 still works over 40 years after it was made.

The illustration on the data sheet did not make apparent a feature of the electrode design —a change in line direction between the segment electrode leads and the backplane, which allowed the display to tolerate slight misalignment of the front and back electrode patterns. This electrode design approach was a first for TN-LCDs. The electrode layout did not solve the problem of little bumps which appeared at the edges of segments when the front and back planes were not precisely aligned. Early DSM watchmakers did not have this problem because they used precision photoetching, whereas Ilixco used screen printers to create electrode patterns.

BUILDING THE FIRST TN-LCD WATCH

In the summer of 1971, Jim decided that Tom should attempt to build a fourth prototype, a functioning TN-LCD watch. His goals were to have a demonstration device that would prove the advantage of the extremely low power drain of their field effect display as well as its ability to be multiplexed. Jim also wanted to show that a liquid crystal display that did not require an auxiliary light source could be made. Finally, he hoped that a watch prototype would show that a TN-LCD was the ideal replacement for the "you-have-to-push-a-button-to-see-the-time" LED display used in the widely advertised Hamilton Pulsar watch.

It took three months for Tom to build this watch, using only the modest equipment at their East Main Street office. From the previous prototype displays they had built, Jim and Tom learned that the laboratory grade polarizers were too dark and much brighter polarizers were needed. The calculator display used the same polarizers as those in sunglasses, which were also too dark for reflector displays. Ilixco's friends at the Polaroid Corporation sent samples of an experimental polarizer that had a very high light transmission of 55 percent. Because it exceeded the theoretical limit of 50 percent, a great deal of light was lost when two of these polarizers were crossed. A pair of them looked light blue.

Ilixco needed a bright white reflector to mount behind the rear polarizer and began searching for the whitest possible reflecting surface. They tried white paper, milk glass, ceiling paint, ceramic plates, optical standard aluminum oxide and even Day-Glo florescent paints in various colors, but the brightness of these surfaces was very low, producing a dull, dark display background. The paints proved especially disappointing because color does not improve brightness and the polarizers absorbed the ultraviolet light that is needed to cause fluorescence. They tried very light blue polarizers, but there wasn't much improvement.

Jim and Tom set aside the problem and took the prototype watch to Bulova Corp. headquarters in New York City. Jim was always very good at arranging to meet division managers, directors and presidents and showed the prototype to Bulova's

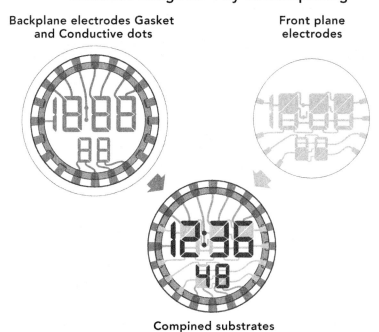

FIGURE 7. TN-LCD Design of the watch prototype for Bulova

chief engineer. Bulova had a venerable history in watchmaking and watch design. In 1960, they introduced the Accutron watch, which used one circuit utilizing a single transistor to power a tuning fork (360 Hertz) that drove the mechanical gear chain. They advertised the Accutron as the first electronic watch, but with only a single transistor, it certainly wasn't an all-electronic watch like the Ilixco prototype. The Accutron's successor, the Accuquartz, had a battery operated crystal that vibrated at a high frequency and was divided into smaller signals that drove the motor in the watch. Ilixco thought that Bulova might regard the all-electronic prototype as an improvement.

Bulova's chief engineer was impressed and showed the prototype to the dozen engineers working under him. Most of them were mechanical, not electrical, engineers, so there was a culture gap. After an hour, the chief called Bulova's president, who looked at the prototype and said nothing about it. But the chief engineer was a fan of the popular comic book detective Dick Tracy and especially Tracy's watch, a two-way wrist radio that the detective talked into. As the Ilixco partners were leaving, the chief engineer gave them reprints of an article he had authored for a watch industry trade journal with his prediction that someday people would wear wrist watch communicators like that worn by Dick Tracy and Ilixco's prototype reminded him of Dick Tracy's watch.

Bulova's president didn't see the potential of the LCD watch and told Jim weeks after the demonstration that he wasn't interested in it. Another American watch company, Hamilton, was making an all-electronic watch at the time, the first to use an LED display. Ilixco didn't consider approaching them because they felt that an LCD alternative would be threatening. In addition, the Hamilton watch was jewelry-like in appearance with its sparkling red LEDs. In contrast, the LCD looked dark and dull.

From building the calculator prototype and other laboratory experiences, Tom learned that a liquid crystal display must be driven by an alternating current (AC) signal. The calculator drive that had been applied to Nixie tubes supplied direct current (DC) to the display, and an LCD could only tolerate that for a short time because of degradation. Therefore, Tom built the circuits from the new CMOS logic devices made by RCA. He was able to create an AC LCD drive scheme that would ensure a long operating life. This was very important, as clocks operate 24 hours a day, or 8000 hours each year, longer than the lifetime of a typical automobile.

By the time Ilixco built the watch, they had thought about how a liquid crystal display differs from electronic devices such as semi-conductors and LEDs. A liquid crystal display is a kind of electro-static mechanical device instead of an atomic-electronic one. The complete watch prototype the company built was very close to the actual TN watch they made a few years later.

A few months after the trip to Bulova headquarters, Jim and Tom looked at the watch display and wondered why the light blue polarizers had not brightened up the display. Jim had an *aha!* moment and realized that every reflector they had tried was depolarizing.

Armed with this realization, they found a linear polarizer sample and placed it over various materials that were handy, such as bright white paper. They observed that light that passes through a polarizer and is reflected by a piece of paper is pretty bright when viewed directly. However, when incident light passes through a polarizer, the light that is reflected off the paper is depolarized. This reduces the amount of light reflected back through the polarizer.

Jim knew that light reflected from a polished metal surface was not depolarized, and placed the polarizer sample over an aluminum mirror. The reflected light was twice as bright. However, an LCD with a mirror behind it would only look good if the observer was in a direct line to the reflected light, and the location of that line was determined by the direction of the incident light. In other words, a display would look great if the wearer held up their arm to catch the sun coming over their shoulder and tipped the watch face just right so as to beam the reflection into their eyes making it difficult to see the display.

Jim and Tom needed a reflector that diffused light and thought a ground glass surface would work. Jim was now able to pinpoint the requirements for a suitable reflector: the material needed to be highly reflective yet non-specular, and the polarization of light had to be preserved upon reflection from the material. Tom took their mirror sample home and ground a diffusing surface on the side opposite the aluminum coating. The grinding of many tiny grooves in the glass surface diffused light by a combination of reflection and refraction, processes that do not alter polarization. He and Jim tried it out behind one of the displays they had built. It worked beautifully. The reflective display they were able to assemble can hardly be distinguished from simple reflective LCDs seen today.

To that success, they added another important feature: an efficient back-light. As an electronics tech, Tom had used oscilloscopes for some time. The older ones used a piece of clear plastic over the oscilloscope tube face that has a grid scratched into the rear that acts as a graticule (grid with x-y

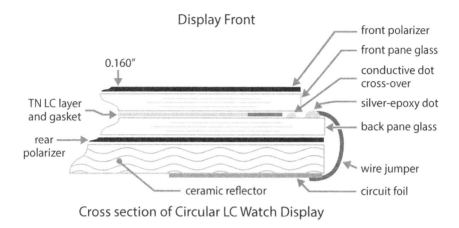

FIGURE 8. TN-LCD cross section

coordinates) that can be used to make waveform measurements. This graticule is illuminated by lamps that are hidden at the edges behind a cover. Tom realized that this scheme could be applied to the non-depolarizing reflector. Jim loved this idea, saying, "If we also aluminize all four of the edges (an easy thing to do), then the light will be trapped inside and the reflector will act like a light box." Light in this light box that came from lamps at the edges would bounce around, and the only means of escape would be through the ground light scattering surface in the front.

This search for a non-depolarizing reflector resulted in a patent for a reflection system for liquid crystal displays, and is an integral part of reflective LCDs in watches, calculators and similar devices even today. Ilixco also introduced an edge light which allowed the display to be illuminated at night. This configuration is still routinely used on LCD watches.

The invention of the special reflector with edge lighting elegantly solved all of the problems of using LCDs in watches.

LEDs drew too much current to operate continuously, but when a button was pressed, they lit up and could be seen in the dark. The TN-LCD drew such a tiny amount of power that it could operate 24 hours a day for several years, and when a button was pressed, the edge light could illuminate the display from the rear so the TN-LCD could be seen in the dark. The TN-LCD had another advantage over LEDs; it was visible outdoors in the brightest sunlight.

12. MAKING A MANUFACTURABLE WATCH DISPLAY

"You start out by discovering the basic principles of liquid crystals, and once you understand them, they represent a scientific base. Your understanding puts you ahead of everybody else, and maybe you understand them in a different way from other people.[...] There was a lot of talk about watches that needed displays. We knew there was a need for displays, and we acted on it."

—Jim Fergason

THE KENT FACILITY had become far too small for Ilixco so Jim took advantage of a real estate connection of Jim Bell's and moved the company into 10,000 square feet of warehouse space in September 1971. There the partners set a new goal of making a manufacturable field effect display and electronic drivers. The warehouse was located in a brand new industrial area at 26101 Miles Road in Warrensville, a suburb southeast of Cleveland, Ohio. Ilixco eventually rented space in a building five minutes away from the Miles Road facility and put in a pilot plant for manufacturing liquid crystal materials.

To fund the expansion, new investors and customers were needed. Jim had previously met Seiko's technical guru, Masatoshi Tohyama and considered Seiko as a perspective investor. The Japanese watchmaker was interested in Ilixco's display products and sent the engineering manager of their electronic watch group to the new facility. Ilixco promised to fabricate sample TN-LCDs for Seiko. Kintaro Hattori had founded the company that became Seiko in 1881, and his descendants still owned and ran it. Jim traveled with Jim Bell to New York City to meet Ichiro Hattori, one of the Kintaro Hattori's grandsons, at the Princeton Club.

Following that meeting, Jim Fergason made the first of several trips to Japan. The Japanese treat guests with great hospitality, and Seiko proved a wonderful host. The engineering manager met Jim at the train station in Tokyo. The manager easily spotted Jim because he was the tallest person on the platform. A driver took them to the exclusive Imperial Hotel and then to dinner at the fanciest French restaurant on the Ginza. Jim also toured the Imperial Palace grounds and Meiji Shrine. He later stayed at a traditional Japanese inn, where he slept on a futon on the tatami mat floor.

The manager took Jim to a meeting with Kentarō Hattori, the oldest grandson of the company's founder and chairperson of all of the companies in the Seiko group. Ichiro Hattori, who Jim had previously met in New York, also attended. The most vivid memory Jim had of the meeting was of the manager falling to his knees when he entered the room where

Hattori was sitting, and pressing his head to the floor in a gesture of deep respect.

Jim was later told that Seiko had decided not to pursue investment in Ilixco because Hoffmann-La Roche had filed a competing patent for the TN-LCD in the U.S, and it had not yet been determined who owned the patent. Given this, Seiko was unclear how risky an investment would be. Jim was disappointed. In retrospect he felt that he should have pressed Seiko harder to make a deal because the backing might have changed Ilixco's fate.

Jim brought a sample TN-LCD with him and demonstrated it for Masatoshi Tohyama of Seiko. At that time, Tohyama was researching RCA's dynamic scattering mode displays, but the superior image quality of the TN display instantly impressed him. The image was "as clear as if it were ink on paper." TN displays also had low voltage requirements and a longer operating lifetime.

Jim gave Tohyama the sample display to keep, and Tohyama switched Seiko's research from DSM displays to "field effect mode" displays. He also decided that Seiko would undertake in-house development of displays rather than buy them from foreign companies. Tom Harsch pointed out that anyone who received the invention would have found it easy to study how it was made and duplicate it. The only element that would not have been easy to determine was how Ilixco oriented the liquid crystals to the inner surfaces of the display.

Jim courted another Japanese investor, Osaka-based Omron Tateisi Electronics Company, which was owned by Kazuma Tateishi. There are social requirements to doing business in Japan, and drinking is one of them. Nobuo Tateisi, one of the founder's sons, took Jim on a jaunt along the Ginza one evening. Nobuo knew every bar, and everyone in every bar knew him. The president of another American company joined in the fun. Late in the evening, Nobuo picked up several bar hostesses and invited everyone to his ski cabin up in the mountains. Jim had had enough entertainment and decided not to go. The other executive from the U.S. later wished he had made the same decision because he slipped in the snow and broke his leg outside Nobuo's cabin.

The socializing paid off. Jim negotiated a deal with Omron's president to invest $1.5 million in Ilixco for a 10 percent ownership position. Omron told Jim that they eventually planned to manufacture the displays they needed for their equipment. This led to a license agreement. As part of it, Ilixco committed to teaching Omron personnel how to manufacture TN-LCDs.

Omron celebrated the investment deal at a traditional Japanese restaurant, with geishas in attendance. Omron took a photo of Jim at the restaurant surrounded by the geishas in their kimonos and mailed it to Jim after he returned home. An employee hung this photo in the company lunchroom to remind everyone of Jim's "hard work" in Japan.

GROWING THE COMPANY

By February 1972, Ilixco was producing several displays a week which they delivered to Compar, their distributor. Initially, the means they used to create these displays were inadequate, and the displays only lasted 6 months. Employee Karen McDonald remembered Jim saying to her one day, "All I want is 300 good displays a day from you." Karen remarked that those were the days when the company was getting maybe 2 displays out of 20 that actually worked without shorting out or having lost segments on a digit.

The staff then consisted of 20 employees, most from Kent and LCI, although several, including an experienced, enthusiastic chemical technician named Martha Dattilo, were from Cleveland. Everyone was young, excited, and focused on the company's goals.

Jim was bubbling with ideas. Duane Werth, a psychology department tech who worked at LCI and then at Ilixco, remembered Jim coming to the lab to talk about an idea he had. He'd explain the issue and talk about possible solutions. As he got to the second or third solution, he'd begin walking out and talking as he left. All of their conversations ended with Duane looking at Jim's back as he walked out the door. Duane knew Jim wasn't being rude, but had hundreds of thoughts and ideas in his head and was on his way to the next conversation, heading to someone else's office to relay those new ideas.

Jim didn't plan to pursue the research he had begun at LCI on the medical aspects of cholesterics because Ilixco didn't have the time or money to obtain FDA approval for any products that might be developed. However, they started a subsidiary called Liquid Crystal Biosystems, headed by Tom Davison, that sold a product called the Lixkit, a set of four different cholesteric liquids in aerosol spray cans, intended for medical applications involving thermography and temperature measurement. The kit also contained a new black skin paint that was formulated for aerosol spray and easier removal with soap and water.

Liquid Crystal Biosystems also developed a highly elastic and flexible liquid crystal film composition and the methods to produce it. The liquid crystal film was composed of liquid crystal and black layers with a medical grade adhesive. It was much easier to use and less cumbersome than the spray. The last product they developed was the first hand-held battery-operated electrocardiogram (EKG) that was enabled by liquid crystal display technology. The dot matrix display prototype was produced in the Ilixco lab and the electronics were developed by Bruce Taylor of the Akron City Hospital Vascular Research Lab.

Jim boldly pushed production, and the company went from producing several displays at a time to 10 to 20 displays. He and his partners began perfecting a solution to each problem they encountered in making a manufacturable display. Jim attributed his success in solving these issues to his strategy of considering his inventions as an entire system. In this, he followed in the tradition of Thomas Edison, who invented not just devices, but entire systems for using them.

NEW SOLUTIONS TO TN-LCD CHALLENGES OF SURFACE ALIGNMENT

Eliminating the undesirable visual artifacts in the images produced by TN-LCD displays was an early challenge. Jim found the problem involved several related issues. The image produced by the early TN-LCDs typically consisted of dark, energized characters or figures on a bright, unenergized background. For the display to be legible, the background needed to be as bright as possible and the characters as dark as possible. From some directions the characters were very dark and the display highly legible, while from others the characters were lighter and the display harder to see. This meant that there was a best viewing direction. Second, the energized areas of the display would sometimes look mottled instead of uniformly dark. Both the dark and light areas in the mottled pattern varied in brightness with viewing direction.

Once the partners had achieved sufficient insight into the optics of the TN-LCD, they designed the twisted nematic structure so that they could set the orientation to the best viewing direction, which was 45°, corresponding to the clock face orientation from which a watch display is typically viewed. This darkened the digits, improving the contrast and making the display easier to view.

Two related issues called reverse tilt and reverse twist were found to cause mottling. When a display was turned on, the liquid crystals untwisted. When the display was switched off, the liquid crystals twisted back through 90 degrees. Some molecules twisted back to the right and some to the left, causing the mottled appearance. This was reverse twist. Jim's explanation of this effect was that the tilt angle of liquid crystal molecules at the surfaces determines the direction of rotation of the twist.

The simplified explanation of reverse tilt is that even if the entire display area had uniform twist in one direction, e.g., was right-handed, sometimes the applied field would "lift" the molecules uniformly in the central layer. In some areas the molecules lifted in a way that caused right-handedness, while in others the lift caused left-handedness. The result is that off-axis, the display showed patches of unequal optical density.

Fixing the problem of reverse tilt and reverse twist had to wait until Jim developed a reliable surface alignment means.

Jim knew that surface alignment was going to be important from the very beginning, because he had seen through the polarizing microscope that the twisted nematic optical effect was precisely controlled by the surfaces. Good surface alignment in a TN device is uniform, homogenous orientation of the liquid crystal molecules that are next to the inner surfaces of the cell. It would also establish a 90-degree twist between the two surfaces and provide the maximum and minimum transmissions and best contrast ratio.

The first observations on alignment had been made by early pioneers in liquid crystal technology, who found that liquid crystals would align on glass substrate in different ways, but the results were inconsistent and not always reproducible. The scientific basis of alignment was not understood and could, at best, be described as an imperfect art. It was known that alignment on the substrate affected the alignment within the bulk.

The older techniques for producing alignment included chemical cleaning or etching, and rubbing the surfaces with horsehair, leather, paper and rouge. The alignment techniques Ilixco tried first were not uniform over large areas and only lasted for several days.

It ended up taking five years for Jim to solve the surface alignment problem from start to finish. There were many reasons for this. First, the mechanism that causes the liquid crystals to align parallel had to be uniform. The molecules had to be tilted up slightly from the surface at a uniform angle. Because of reverse twist, Jim realized that the liquid crystal molecules could not be flattened to the surface. Also, during production, the workers needed to be able to align the surfaces of displays with etched electrodes. The polymer alignment coating that needed to be developed had to mask the uneven influence on tilt of two different atomic surfaces: glass and indium-tin oxide. The alignment material and technique had to be reasonably easy to apply—the company couldn't afford to make liquid crystal displays using a process that required vacuum chambers. Finally, they needed an alignment coating that the liquid crystal could not dissolve and that moisture entering the cell would not alter. These requirements would have been easy to meet if the surface was an object of ordinary dimensions, but it was an ultra-thin layer of just a few molecules.

Jim had initiated studies of normal alignment when he had arrived at LCI because alignment was needed to construct the devices to study electric field effects in highly resistive, negative dielectric materials. He, Ted and Tom had used many of these older techniques to produce both parallel and normal alignment in cells they made to study dynamic scattering. The performance of dynamic scattering wasn't particularly improved by these techniques. However, de-ionized liquid crystals didn't scatter, and that permitted experiments that were much more controlled. So Jim explored how electric fields altered the structure of negative nematics with a director oriented (without a field) normal and parallel within the cell. One of these experiments produced unsurprising results: a nematic liquid crystal with negative dielectric anisotropy that is aligned parallel to the surface by rubbing retains that alignment when an electric field is applied.

Jim continued to gather clues from widely scattered sources to determine how to align a liquid crystal to the display surface. He read previous reports by German and Austrian scientists. He used knowledge gained from experiments he had conducted at LCI to align dynamic scattering and other negative nematic and untwisted field effect devices. Additionally, he obtained clues from his friends in the Polarizer division of Polaroid and the polarizer patents of Dr. Edwin H. Land. Finally, he, Ted and Tom read books on the nascent science of surface chemistry, and learned practical ways to analyze the cleanliness of glass. However, nothing they read helped them understand liquid crystals and "the rub."

This work on alignment resulted in the development of a procedure to identify suitable parallel alignment materials (a procedure that he eventually patented).[31] The procedure began with ultrasonically cleaning the glass, rinsing it in distilled water and then drying. The glass was next rubbed against paper and the cell assembled by squeezing a thin film of liquid crystal between the glass substrates. The substrates were separated by between .00635 and .0254 millimeters by two thin strips of Mylar TM polyester of the proper thickness (Dupont provided all of the spacers at no charge.) The microscopic uniformity of each cell was examined under the polarizing microscope. Jim could see some of the causes of lightness or misalignment, like dust specks distorting the director. This process became routine for evaluating various rubbing materials, surface coatings and rubbing devices.

Using this procedure, the company investigated a wide range of rubbing materials. Dora supplied many of these materials, including polyester, silk, cotton and velvet. For several months, a piece of velvet was the preferred material for producing parallel alignment, with the backing glued to a board to serve as an edge guide. The fabric was dampened with a solution of various materials.

Velvet produced results that were initially acceptable, but the alignment was found to degrade over time. This was undoubtedly due to the small amount of material deposited on the glass. The velvet also had a feature that Jim characterized as "bite," referring to its interaction with the surface being rubbed. The next step was to apply a coating directly from the solution onto the substrate, with the thought that this would provide the rubbing material with a second material that would "take" the rub.

One of the first materials used as this alignment coating were the natural oils on the assemblers' hands. The partners had observed that the quality of alignment was better when the assembler accidentally touched the velvet with sweaty hands before rubbing the glass. Alignment produced in this way also had a short lifetime, degrading after only several hours. The degradation was believed to be caused by the liquid crystal dissolving the coating. However, this result was encouraging because it suggested that if the right alignment material could be found, a microscopically uniform and durable alignment could be achieved. Some initial success was obtained with polyethylene glycol, and this material was used in early TN-LCD fabrication.

Jim observed that when the clean glass was rubbed with a paper towel from LCI's men's room, the alignment was excellent. He realized that in addition to wood pulp and clay, these cheap paper towels contained PolyVinyl Alcohol

(PVA) as sizing. He also knew that Edwin Land had rubbed and treated PVA with iodine, and the iodine then showed alignment. These experimental observations were bolstered by Oseen's theoretical work pointing out that solids would also have directors, but the small associated energies would be completely masked by elastic properties. All this evidence suggested that PVA might be a good parallel alignment material. Jim had already used the material to encapsulate temperature-sensitive cholesteric droplets into films, and knew that it was not soluble in liquid crystals. Tom bought and read a new book, *Poly (Vinyl Alcohol): Basic Properties and Uses*, which was *the* book on PVA and remained so for three decades. The research librarians at the Cleveland Public Library also sent a thick stack of polarizer patents for Jim and Tom to study.

Jim obtained samples of commercially available PVA, but it had too low of a resistivity because of the way it was synthesized. PVA is made by reacting vinyl acetate to form polyvinyl acetate. It is then treated with sodium hydroxide to remove the acetate groups, leaving an alcohol group. The product, as typically sold, may have as much as 10 percent residual sodium acetate. For use with liquid crystals, it would need to be purified.

Sardari set up a device called a "soxlet extractor" to remove the sodium acetate by bathing the PVA in warm pure methanol. Twenty-four hours later, he obtained enough ultrapure PVA to use in production for a week.

The ultrapure material provided the dramatic improvement in surface alignment that the company had been searching for. Now there were no more "light" or "heavy" rubs. The displays had a uniform alignment and tilt angle that always eliminated reverse twist, and a uniform, constant, and quick turn-off. With purified PVA in hand, Jim developed a process to utilize the PVA. He tried casting a thin film of PVA onto the substrate from an aqueous solution. After drying, the film was rubbed in one direction. The result was a rubbed thin film of PVA-reliably aligned liquid crystal materials with a low tilt angle, independent of the rubbing material.

As hoped, when the layer of liquid crystal was uniformly aligned, the problem of haze was eliminated and the cells did not scatter light. Other alignment materials were later found and tested, and it became possible to identify the material characteristics responsible for producing good surface alignment. Jim's analysis of surface orientation had not produced the kind of knowledge that could be published in a scientific paper, but it was reliable, thorough, and practical enough to develop excellent surface alignment techniques.

To perform the uniaxial rubbing process, Ilixco next had to develop a rubbing machine. When they were trying to determine the buffing speed, they found that as the operator held the glass substrate against the wheel, the high speed rotation burned the operator's fingers. They were concerned that the friction generated by high speed rotation would burn the alignment coating. Occasionally, a corner of the glass substrate got caught by the wheel, tearing it from the operator's hand and slamming the glass onto the tabletop. A slower speed made this less likely, and, when it did happen, less dangerous. Jim brought in the Sears radial arm saw sitting in his garage shop to use for production. It allowed the glass to be held on a flat surface and aligned at any direction with respect to the edge. The saw blade was replaced by a stack of cotton buffing pads. Since the saw had a universal motor, the tangential velocity could be controlled by a variable transformer.

Although test cells that incorporated an aligned layer had been fabricated for many years, Ilixco was the first company to produce liquid crystal displays of any type that included two aligned layers in a reproducible, controlled manner. Jim remarked that the development of the first commercial alignment layer "was as important to the success of the TN-LCD as the twist itself." Despite its dominance, the TN-LCD is just one of many liquid crystal electro-optic effects today. However, every liquid crystal electro-optic effect requires some form of alignment layer, and all such alignment layers derive from Ilixco's initial work.

SOLVING THE REVERSE TILT PROBLEM

Having developed a reliable surface alignment means that solved the reverse twist problem, the company was able to work on solving the reverse tilt problem. One of the best industrial researchers at the time, E.P. Raynes, was also researching reverse twist and reverse tilt. In a 1973 paper, he analyzed reverse twist and its structural causes, but didn't offer any surface preparation solutions to the problem. A year later, he sent a second paper to Jim describing a method for solving reverse tilt.

Raynes' solution was to add a sympathetic twisting or chiral agent to the nematic liquid crystal, a clever solution because it harmonizes the rubbing direction with the twist direction. This was the solution Ilixco adopted. Jim had developed the means to synthesize high purity cholesteryl chloride at Westinghouse. Ilixco produced this material, and the chemists doped the material with this chiral agent. They blended cholesteric into the liquid crystal to produce a twist bias of 1 to 2 degrees. Jim had calculated the amount based on the inherent twisting power and the mole percent of cholesteryl chloride. Even today, this solution is universally applied.

13. THE "TIME COMPUTER YOU WEAR ON YOUR WRIST"

> "Being an inventor is a simple concept. Being a successful inventor and earning a living is not so simple. It requires motivation, opportunity and luck."
>
> —Jim Fergason

OVER A 10-MONTH period between 1972 and 1973, Ilixco grew to 200 employees and increased production to 1,000 watch displays a month for the watch company Gruen. Most of these employees worked in Sardari Arora's brand new 1,000 square foot chemical lab for liquid crystal synthesis. Display production was carried out in two "clean rooms" the company constructed behind the lab.

Tom had read an article in *Electronics* magazine about how graphics helped relieve boredom in clean rooms, so he decorated the walls with large, bold, green and white graphics. There was an airlock, and positive air pressure supplied through high efficiency particulate air filtration and activated charcoal filters mounted in the ceilings. The production equipment consisted of workbenches and a large screen printer of the type used to decorate T-shirts. The workers used this printer to print display transparent electrode patterns and the spacer-seal that went around the substrate perimeter.

Manufacturing facilities at Ilixco, 1974

The company created their own liquid crystal material for the displays. Sardari was in charge of producing the materials, but Jim gave valuable input. Although he wasn't a chemist, he had acquired by now a wealth of experience, insights and knowledge about liquid crystal formulations from his time at Westinghouse and LCI. It was the "kind of knowledge and experience you couldn't get out of a book," said Ken Marshall, one of the Ilixco chemists. "He knew an amazing amount about liquid crystal structures and structure-property relationships. He knew what basic shapes of molecules would create what properties. Back in those days you had to have physicists and chemists working together to figure that out, but Jim knew it. He knew more than Sardari Arora."

Developing a liquid crystal material for the display was very difficult because the material had to meet a long list of competing chemical, electro-optical and physical requirements. The final mixture also needed to be liquid crystal over a broad ambient temperature range.

The approach Jim and Sardari took to creating the material was to blend liquid crystal and non-liquid crystal materials that had the necessary properties. Sardari started by using a stable, Schiff base nematic that had a negative anisotropy. This material was a liquid crystal at room temperature but only over a narrow temperature range. He added a cyano derivative nematic liquid crystal that was a solid at room temperature, but that had a strong positive anisotropy. The positive material had only limited solubility in the negative material. To address this problem, a non-liquid crystal material with a molecular shape similar to the liquid crystal molecules was added to enhance the solubility of the positive material. Finally, another nematic liquid crystal material was added that had a melting point of about 40°C, but that was in the liquid crystal phase over a very wide temperature range. Determining the proper proportions of these four materials took two weeks and over 100 trials.

Dynamic scattering devices tended to have a short lifetime because the current flow caused a voltage drop at the electrodes used to form the digits, and the digits underwent an electrochemical reaction and turned opaque. The TN-LCD required 10 times less power than a dynamic scattering device, which reduced the decomposition of the ITO electrodes, extending the display lifetime. (Jim used to say that "the amount of power needed for a TN-LCD device was that same as a person walking over a wool rug in their stocking feet.)

In order to prevent current flow in the field effect device, it was also necessary to reduce ionic contamination by purifying the liquid crystal material. Jim remembered the unpublished trick of the trade that he learned from the chemists at Westinghouse using Fuller's earth materials, which have a vast number of active sites and a large surface area that adsorbs ionic impurities. The first step in the purification procedure, activating the Fuller's earth, required heating it to a temperature of 220°C to drive existing contaminates out of the material. The nematic liquid crystal was then stirred in and the resulting slurry centrifuged, trapping the active impurities in a near solid layer. The liquid crystal was decanted and filtered through a Millipore filter to remove particulates. The liquid crystal that passed through the filter was purified. The resistivity of the resulting liquid crystal was very high, on the order of 10^{15} Ohm-cm. One of the advantages of this procedure is that it could be used as the last step in formulating mixtures, and no further handling of the purified material was required before incorporation in the display.

Jim also used Zeolite clay for purification, a product sold by Union Carbide that came in different pore sizes. Later, he used a mixture of Fuller's earth and Zeolite clay to absorb the anions (negatively charged ions), cations (positively charged ions), and small molecule contaminants from the liquid crystal.

Jim and Sardari successfully developed a Schiff base liquid crystal blend that met the requirements of these first generation TN-LCDs. But Ilixco's customers claimed that these materials had insufficient environmental stability, meaning when they got heated or frozen, they stopped working and never worked again. In response, Sardari developed ester-based materials that were even more thermally stable. (U.S. patent 4,086,002 resulted from this work, with Sardari Arora as the inventor.) The mixture had a lower threshold voltage and excellent performance. Sardari was a perfectionist when it came to producing quality materials. His products were as pure as they could be, and he had an extraordinary ability to train staff in making them.

Undertaking the in-house development of materials allowed the company to explore a wide range of material properties and to be experiment extensively in process development. Ilixco eventually developed sufficient material design and synthesis capabilities to tailor make the material for direct drive or multiplexing applications. The multiplexable materials used compounds called benzalazines, which were very stable and excellent for lowering the threshold voltage, although they imparted a strong yellow color to the mixtures that the customers disliked.

The chemists also synthesized a low viscosity, room-temperature liquid crystal material based on methylbenzyl-butyl aniline called MeBBA (not to be confused with the better known liquid crystal material that chemist John Dreyer developed called methoxybenzyl-butyl aniline, designated with a similar acronym, MBBA.) The company's dream was to manufacture these materials at the multi-kilogram level and sell them, but this dream was never realized. All of the materials were used internally.

FORMING THE DISPLAY'S GLASS ENVELOPE AND SEALER

The next manufacturing-related task addressed the dual challenge of developing the means to form the glass envelope and fill the resulting cell. For the solution, Jim drew on the practical lab experience he had filling flexible cholesteric liquid crystal devices at Westinghouse.

The diagram in Figure 9 on the next spread illustrates the core production process determined by Jim's decision to use polymer as an alignment agent as well as a seal to contain the liquid crystal and exclude contaminants. The process was called a "drop-fill" technique, because after a small quantity of the liquid crystal display was pipetted onto the substrate that lay in front of the assembly, another substrate was brought down slowly into contact with the liquid crystal and dropped. The volume of liquid crystal material dispensed onto the bottom substrate was carefully measured so that when it spread, it would just fill the TN-LCD cell.

This technique was efficient and economical, and it also mitigated a second fill-related issue that Jim had observed. As a liquid crystal flowed across an alignment surface, various components in the liquid crystal mixture tended to sequentially "plate out" onto the surfaces. While this same plating effect is the basis of liquid chromatography, it resulted in visible, non-uniformities in the finished TN-LCD. The use of drop-fill minimized the flow path and liquid crystal plating.

A consistent method to seal the two substrates together also needed to be developed. The technique chosen was to screen

1. Begin with glass plates coated with an indium oxide electrically conductive transparent coating.

2. Screen print acid resistant ink in the form of electrode patterns.

3. Etch away the indium oxide coating to create the electrodes, strip away the ink, and clean the glasses.

4. Screen print the polymer seal-spacer onto cach glass.

5. Apply the conductive dots and bake the pieces to drive the solvent from the seal and the dots.

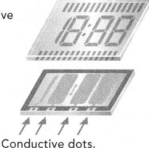

Conductive dots.

6. Apply the LC alignment coating and buff, at 45° to the display and 90° one to the other, in order to orient the "dark quadrant" toward the viewer.

7. Drop the LC onto the bottom plate, lower the top plate in order to spread out the LC so that it fills the cell and oozes out past the seal-spacer. Clamp the plates together.

8. Heat the clamped assembly to soften the dots and Knit the seal.

9. Apply front and rear polarizers.

DONE

FIGURE 9. TN-LCD manufacturing process developed at Ilixco

print a gasket around the perimeter of one or both substrates. The two substrates, along with the liquid crystal contained between them, were then heated to seal the cell. The trick in implementing this process was to find or develop a seal material that would adhere to the substrates in the presence of the liquid crystal. The amount of liquid crystal inside the display was tiny, but without an effective seal to keep it inside the display and moisture out of the display, the display would deteriorate.

It occurred to Jim that the problem of sealing the TN-LCD cell was similar to that in certain food packaging applications, where an adhesive that could self-adhere in the presence of animal fats was required. Ilixco began investigating a class of adhesives used in food packaging applications. Jim's intuition was that a plastic seal would be the best solution, but when he tried the silicone adhesive used in food packing, it remained soft. Displays made with it were easily split apart with a razor blade. The silicone adhesive was tacky when it was being printed, and thousands of tiny strings formed when the screen pulled away from the glass. These strings fell onto the glass in the display area. An additional problem was that the liquid crystal slowly dissolved the adhesive.

The company looked for an expert in adhesives to help. In late summer of 1972, they found Harry King, a brilliant independent consultant who had developed an adhesive that bonded certain wing components together in a Boeing jet. Jim asked King to find a thermo-plastic material that would be compatible with LCDs. It had to absorb the liquid crystal that would remain between the two seals after the glasses came together and yet not react with the liquid crystal and destroy its optical properties, causing "blooms" in the cell display area. It also had to create an active moisture barrier, be strong, and form a good bond to the display glass. Finally, it had to be screen printable and form a flat uniform seal-spacer of the proper thickness.

King selected a dozen candidate materials that met the first requirement. After testing these, Ilixco chose a Bakelite (phenolic) resin made by Union Carbide. Meeting the other requirements involved experimentation and testing. Tom Harsch gleaned critical information from six polymer textbooks, and over the next two years, made 2000 test displays that were sealed with phenolic resin. The chemistry tech Martha Dattilo mixed many different formulations of the seal material. The company's production of displays was an ongoing test. Ilixco ultimately produced a seal-spacer that Martha named the "FieldStick," and that proved superior to the best glass "frit" (the glass-to-glass seal used by RCA or Optel for their dynamic scattering mode LCDs).

The seal was so strong that when the watch display was mounted in a machinist vise and the vise tightened, the glass cracked before the seal gave way. Under humidity tests, the seal proved an excellent moisture barrier. This seal-spacer process turned out to be suitable not only for watch displays, but large displays. The gasket materials and sealing processes were later patented by Jim.[32]

It has been over 40 years since Ilixco developed the materials and processes to make TN-LCDs manufacturable. The same techniques they developed to control the best viewing quadrant and remove reverse twist and reverse tilt are still in use today. After years of using other means, the manufacturers of large LCDs are again utilizing, with improvements, the drop-fill technique.

ADHERING THE POLARIZERS

The company next had to figure out how to make the polarizers adhere to the glass cell of the TN-LCD. The surfaces of the glass cell and polarizer were all sources of reflection, and thus of light loss. For the devices to present a quality image these reflections had to be suppressed by optically coupling the polarizers to the glass using a non-birefringent adhesive. Ilixco hired Morgan Adhesives in Stow, Ohio, to custom make a free-standing adhesive laminated between two release sheets, each of which had different release factors.

The material worked perfectly. Ilixco only needed 400 square feet of it, enough for six months of use and the production of either 10,000 instrument-sized displays or 40,000 watch-sized displays. Morgan Adhesives made its profits by producing millions of square feet of materials for applications such as shelf paper and optical films for road signs. They wouldn't make much profit producing such a small amount of material, but Jim Bell had helped the company's founder, Burt Morgan, obtain initial financing to start his business. Morgan was grateful and produced the material.

With all of the major problems in manufacturing TN-LCDs resolved, company personnel conducted an electro-optical evaluation and characterization of the devices. They made an accurate determination of the voltage response curve of the TN-LCD, a key electro-optical property that had only been observed visually and reported anecdotally. They also measured the contrast ratio, the capacitance and resistivity, and the frequency response, which is the relationship between the optical response and the frequency of the applied field. In addition, they investigated the response of the TN-LCD to a voltage pulse as a function of cell thickness. They concluded that the delay in response after the application or removal of the voltage was not visible to the human eye and should not be counted as part of the optical response time of the device.

Jim observed that some alignment coating materials and procedures produced better extinction than others. He determined that these differences in darkness were related to the tilt angle. The smaller the angle between the liquid crystal director and plane of the substrate, the darker the extinction. In an experiment in which the voltage was removed and the switching speed measured, it was found that "dark" cells switched off faster than lighter ones.

SALES BEGIN

Ilixco demonstrated their display technology at Applications Night at the 1972 International Liquid Crystal Conference held at Kent State University. Foreign scientists in attendance, including German and Japanese scientists, had a rare opportunity to see TN displays for the first time. Wolfgang Helfrich, who was with Hoffmann-La Roche in Switzerland and whose name was on their TN patent, gave the plenary lecture.

Ilixco was now capable of manufacturing displays, and Jim hired Dick Strong as the sales and marketing manager. Dick had been a Compar executive and had represented Raychem when it was a new company. Dick began aggressively soliciting customers through an effective sales campaign. His brochure included a clever pun on the name Felix, which was short for Field Effect Liquid Crystal but also the name of the famous cartoon cat. Dick put an image of this cat on the data sheet as Ilixco was able to buy the rights to use the Felix trademark from King Features Syndicate, Inc. for very little money.

Dick's sales campaign brought in new customers. Tekelec-Airtronic (Bordeaux, France) purchased TN-LCDs for digital volt meters and the Dickey-John Corporation (Auburn, Il) bought corn moisture meters.

Ilixco also made a watch display for Citizen Watch Company. The substrate, already containing an etched electrode pattern, was supplied by Citizen.

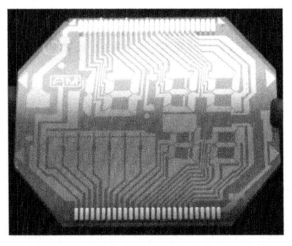

The segment electrode and substrate for the Citizen watch display

Citizen implemented the electrode design without incorporating many of the features Ilixco recognized as necessary for the practical production of TN-LCDs. About a month after delivery of the prototypes, three engineers from Citizen met with Jim, Tom and Sardari. Tom found it odd that none of the many questions they asked him concerned the principles of operation of the TN-LCD, its optics, electrical characteristics or operating life. Instead they discussed the cosmetics of the displays. This cemented Ilixco's realization that watchmakers regard watches as jewelry, and only incidentally as timepieces. Just as the Citizen engineers were about to leave, they made the telling comment that Citizen intended to produce all of the critical materials needed for TN-LCD manufacturing themselves. This shows just how important the watch industry considered TN-LCD-based watches.

THE NEED FOR MORE FINANCES AND A CONTRACT WITH GRUEN

Ilixco began having cash flow problems. In retrospect, Jim felt that in addition to hiring a marketing manager, he should have hired a full-time professional to run company finances rather than continue doing it himself. The accounting firm Arthur Andersen was hired to conduct an evaluation, and concluded that the company needed $5 million to be properly capitalized. An independent expert consulted for a second opinion said $8 million was needed.

Jim asked Jim Bell to raise money. Bell told Jim that the amount of money the company needed could not be raised without unacceptably diluting current investors' stock positions. Instead, he developed a new financing idea—seek out huge orders. All of the contracts up to this point had been for small orders. Bell looked for customers who would place bigger orders and found one in the Gruen Watch Company.

Gruen invited Jim Fergason, Dora, Tom and Jacky to New York. They put them up at a good hotel decorated in the European style, took them to a Broadway show, and wined and dined them at Benihana (the Japanese restaurant where the chefs perform). Jim closed the deal on that trip, signing a contract in the summer of 1972 to supply 4-digit TN-LCDs to Gruen, who wanted to be the first company to introduce a TN-LCD watch for their Teletime line. They planned to sell it during the Christmas season of 1973. This represented the opportunity for Ilixco to move from selling only a few hundred displays a month to selling tens of thousands. Tom was nervous about supplying such a big order because he didn't think the process for making the displays was good enough to give good yields, but Jim jumped at the opportunity.

Ilixco designed the Gruen display in September and gradually ramped up production to 50,000 displays a month. Gruen had the other components of the watches, including the integrated circuits and watch displays, assembled by other contractors. They did not even show Ilixco samples of these items. The earliest Gruen Teletimes made with Ilixco displays were manufactured in 1972 and sold in big cities like New York and Pittsburgh for prices ranging from $150 to $200. Gruen advertised them as a "time computer you wear on your wrist."[33]

To fill the Gruen order, Jim borrowed money from the bank to hire more workers and lease the adjacent 10,000-foot bay in the Miles Road warehouse. The company increased their staff to 200 workers.

FUNNY EMPLOYEE STORIES

Ilixco realized they should have been more careful about who they hired, as Jim recalled many strange employee stories.

For instance, the quality control engineer found the four young Kent State University students he had hired to perform the screen printing operation smoking pot on the roof on their lunch hour, which explained why their productivity deteriorated every afternoon. A shift supervisor almost got stabbed by a woman he had fired. He was escorting her out of the building and she grabbed a box cutter and lunged at him. Luckily, the box cutter was chained to the frame of the door they were going through and the chain didn't extend as far as the supervisor. Another woman was discovered selling sex, using an upstairs storage area for her liaisons. When a handyman found out that his boyfriend, also employed at Ilixco, was one of her customers, he ran through the building smashing the walls with a hammer. A beautiful blonde woman tech worked in the prototype department and was always asking men to do favors for her. One day, Ken Marshall sent his assistant to the solvent shed to get solvent, but when the assistant opened the shed door, he saw the couple on top of one of the drums. He shut the door and came back without the solvent. When Ken asked him where the solvent was, the assistant blushed and said he couldn't tell him. Ken pressed him, and he finally divulged what he'd seen.

"Wait 25 minutes and go back," Ken told him, and sure enough, the couple was gone and Ed was able to get the solvent. Many of the women workers would invite their boyfriends to the plant for a rendezvous in the break room. The night shift supervisor in production, took a tough approach when he entered the break room one night and found a woman with her boyfriend. He yelled at the couple to stop and get out, and the boyfriend said, "Can't I finish?" In response the supervisor whipped out a sidearm and barked, "I said to get out of here." The man ran out of the building.

Vandalism also occurred. The head chemist treated lower level workers imperiously and they sometimes retaliated against him. Several times he forgot to lock his car and found that his car headlights had been turned on to drain the battery. The daytime production manager was difficult to work for and some of the line workers decided to get even with him. After work one day, he got in his car and tried to drive away, but the car wouldn't move. The tires were embedded in to the pavement up to their rims. The line workers had carried out a five gallon can of the solvent chemical DMF (dimethylformamide) and poured it over the tires, causing them to melt into the asphalt. An auto shop unbolted the lug nuts and lifted the car off its tires. Nobody could ever get the tires off the pavement. The tires and lug nuts stuck up until they were finally sawed off at street level. But the workers weren't finished with the manager. Someone poured DMF onto the hood and body of his car, stripping the paint off. Because industrial grade DMF has a fishy odor, the car stank.

AND FRIGHTENING STORIES, TOO

The Ilixco plant was located in Warrensville Heights. Many of the company's workers were women from the surrounding neighborhood. Racial tensions developed between some of the workers. On the second shift one night, many of the women's husbands showed up outside the company building. Fearing a fight, management called the Warrensville police, but they refused to come. The police in the neighboring city of Shaker Heights arrived shortly and told the men to leave.

There were occasional scary incidents in the chemistry lab. One day, Ken Marshall was doing re-crystallization in four liter flasks filled nearly to the top, which sat on four hot plates. The magnetic stirrer stopped, and the solvent superheated from the bottom. The vapors hit the hot plates and ignited. A 14-foot high wall of flame shot from the bench to the ceiling. There were two fire extinguishers in the lab. Ken bolted for the carbon dioxide extinguisher, and Ed Richards, who was on the other side of the room, grabbed the dry powder extinguisher. They both aimed their extinguishers at the fire. It flashed through Ken's mind that the chemicals in the dry powder extinguisher would go everywhere and he yelled "no!" at Ed, but it was too late. Ed's extinguisher went off first, and dry powder sprayed all over the lab and liquid crystal material. It took them three weeks to clean up. Months later, they were still finding dry powder residue in drawers.

All wastes collected from processes were dumped in a 55-gallon drum that stood outside the building. One snowy winter day, a young tech went outside to put waste in the drum. He noticed the top of the drum bulging out and ran back in to get Ken Marshall. The two sped back out and Ken noticed that the top of the drum was as hot as a steam boiler and about to explode. He and the tech grabbed shovels and banked snow on the drum but the snow melted as fast as they shoveled it up. "Go faster," Ken cried out and the two shoveled furiously, finally piling enough snow on the drum so it wouldn't melt. They used a drum wrench to open the drum and let the gas out slowly until it dissipated.

Despite these challenges, Ilixco was still able to rapidly manufacture TN-LCDs and build an excellent technical team. Many of the company's alumni founded or made key contributions to the next generation of American LCD companies. Ilixco also introduced the technology to foreign companies, including Omron in Japan.

Omron sent a three-person technical team to the Warrensville Heights facility for one month in 1973. Tom taught them everything he knew about manufacturing the device, and Ken Marshall everything about making liquid crystal formulas. Ken remembers being surprised when Jim told him not to hold anything back. He doesn't remember Jim telling him about the license agreement he had made with Omron on his 1971 trip to Japan. Omron brought the technology back to Japan and began manufacturing the displays to use in their products.

14. PATENT NUMBER 3,731,986

> "The right of the inventor is a high property;
> it is the fruit of his mind—it belongs to him more than
> any other property—he does not inherit it—he takes
> it by no man's gift—it peculiarly belongs to him,
> and he ought to be protected in the enjoyment of it.
> —Daniel Webster, speech before Congress, 1824

RECALL THAT HOFFMANN-LA Roche filed an application in Switzerland for the twisted nematic idea, claiming Helfrich and Schadt as co-inventors, on December 4, 1970. They were granted the Swiss patent and sometime afterward filed a competing application claiming the invention in the U.S.

On May 8, 1973, after a two-year examination process, the U.S. Patent Office declared Jim Fergason and Ilixco the inventors of the TN-LCD and awarded them the U.S. patent over the competing patent application by Hoffman La-Roche.[34] After Roche was denied the patent, they sent legal and scientific experts to the patent office to try to persuade the patent examiner, Edward Bauer, to change his mind and issue them the patent instead. But Thomas Murray had provided clear and compelling documentation that Jim was the inventor of the device, and Bauer wouldn't budge.

This chapter and several of the next ones continue to tell the story of what happened following that first patent battle—what became a bitter fight for Jim Fergason to defend what was rightfully his invention. As you will learn, the battle becomes far more complex than the original dispute over who owned the patent behind the TN-LCD, and ended up involving numerous parties who all wanted a piece of the potential fortune to be made from what Jim Fergason had created.

THE PATENT BATTLE BEGINS

There were relatively few patent applications for liquid crystal devices in the late 1960s. Edward Bauer was the patent's office top expert on the subject and the patent office assigned him to review both Roche's application and Jim's when they came in late 1970 and early 1971 respectively. Bauer found Jim's attorney John Linkhauer's original application too vague — the U.S. Supreme Court described patent applications as "one of the most difficult legal instruments" to prepare—and sent it back to be rewritten. Jim's law firm's senior attorney Tom Murray filed the rewritten application on April 22, 1971. This was the fundamental TN-LCD patent application. (Tom also wrote four other related patents, all filed in 1973.[35])

The applications reviewed the configuration of the TN-LCD and its principles of operation. They disclosed that

TN-LCDs respond to the root mean square of the applied voltage and described how to multiplex the device, building on the concepts Jim presented in his *Electro Technology* article. The disclosure of practical aspects of implementing a TN-LCD, including discussion of the alignment layer and liquid crystal mixtures appropriate for use in the device, made the applications "enabling," meaning that a person with some prior experience in the field can, by reading the patent, "make and use the full breadth of the invention." The disclosures presented innovative means of making a real, as opposed to theoretical, TN-LCD device.

The United States has long encouraged invention and protected the rights of inventors. George Washington signed the first patent statute into law in 1790, and the United States Patent and Trademark Office (USTPO) was founded as the Patent Office twelve years later. The patent examination process at the USTPO has, for much of its history, been considered equal to or better than that of any other country. It is a comprehensive, professional process, much more rigorous than peer review of scientific articles.

Edward Bauer determined that Jim was the first to invent the TN-LCD based on Jim's *Electro Technology* article and its January, 1970 date of publication. In this article, Jim gave a historical overview of liquid crystal research, and then discussed each optical and electrical property of the nematic and each configurational element needed to create and operate a device. He discussed the birefringent properties of nematic liquid crystals and the ability of thin, aligned layers of such a material to affect the state of polarization of light traveling through the layer. He also explained that placing the nematic between two glass substrates, whose inner surfaces have been prepared by rubbing, aligns the liquid crystal layer. The effects of the liquid crystal layer on the polarization of the light were visualized by placing the device between polarizers which could be either crossed or parallel.

The article then discussed the response of a liquid crystal to an electric field, pointing out that when a liquid crystal has a positive dielectric anisotropy, the long axes of the molecules want to align parallel to the applied electric field. The configurations of the guest host display were reviewed, and a device in which the liquid crystal has a positive dielectric anisotropy and is aligned with its long axes parallel to the plane of the substrates was described. When the electric field is applied, the molecules turn parallel to the field and perpendicular to the substrates. To enhance the contrast ratio of this device, a linear input polarizer can be included.

In "Novel applications use field effect," Jim proposed a novel guest host configuration obtained by "...aligning the liquid crystals parallel to the two plates but with the top plate at right angles to the bottom plate. In the absence of a field, this gives a twisted structure. However, when a field is applied, the materials align parallel to transmit light." By this means, the contrast is enhanced without needing a linear input polarizer.

By reading this article, any liquid crystal device expert would realize that in a twisted nematic guest host device with an input linear polarizer, the dye could be removed and a second linear polarizer added at the output to absorb, or transmit, polarized light in a similar manner as the dye. In effect, the dye has been moved out of the liquid crystal and onto the surface of a plastic film. That is exactly the configuration and principle of operation of the TN-LCD, and the article is the first written document to make this disclosure.

The article also discussed a device configuration in which the liquid crystal has a negative dielectric anisotropy and the molecules are aligned perpendicular to the substrates. This configuration can also be used as the basis of a display, and is today commonly called a Deformation of Aligned Phase (DAP) device.

With publication, the information in this article entered the public domain. As prior art, this meant that the configuration and principles of operation of the TN-LCD could not be patented by anyone other than Jim. An Ilixco internal document dated September, 1970 also documented the configuration and principles of operation of the TN-LCD. This was the original TN-LCD disclosure Jim wrote and later gave to Tom Murray to form the basis of a new patent application, one of the five applications Murray filed in 1973. Jim showed this disclosure in 2002 to Hirohisa Kawamoto, a scientist who was researching for an article he would eventually write, entitled "The History of Liquid Crystal Displays."[36]

In the article, Kawamoto wrote about a meeting he had with Jim in which Jim "showed me a nonpublic document, 'Liquid crystal nonlinear light modulators using electric and magnetic fields.' The document contained a full account of the concepts of the twisted nematic mode." Kawamoto noted that the document stated that "the invention was conceived on December 30, 1969, and explained to Ted R. Taylor and Thomas B. Harsch the same day. The invention was first reduced to practice on April 5, 1970." Taylor and Harsch both countersigned the document and Jim Bell recalled being present when it was signed.

Despite its relevance as the earliest document describing the TN-LCD, Edward Bauer only used the *Electro Technology* article to establish priority. Roche knew about the *Electro Technology* article, but did not, as required by U.S. law, cite it as prior art when they filed their competing application. Years later, Jim met Bauer in person and Bauer told him that that Roche's patent was clearly anticipated by the prior art. By starting with the basic science in the article, any expert could end up with the TN-LCD invention as disclosed. In addition, Roche enclosed no enabling material in their patent application, likely reflecting the lack of laboratory work they had done by the time of filing. Bauer told Jim that their disclosure did not allow anything to be built.

At the time, U.S. patent law did not allow foreign applicants to present evidence that originated from a date prior to the date of their application. However, patent holders were entitled to use evidence that derived prior to the filing date of their patent application, which worked in favor of Fergason. Fair or unfair, this was the law. It did not put Roche at any disadvantage anyway, because the company had no documents substantiating their claim of invention of the TN-LCD that pre-dated their patent application.

Roche was not in a position to manufacture displays. However, they produced chemicals, and chemicals were used to make the displays. Jim had been courting the Japanese for funding, and Japanese companies knew all about the invention. When Seiko and other Japanese companies approached the Swiss giant wanting to do business, Roche recognized the enormous potential profitability of the invention and wanted the U.S. patent for it. According to "Contributions to Today's Liquid Crystal Display Technology," an article by Martin Schadt about the work that he and Helfrich did to develop their version of the TN-LCD, Roche started liquid crystal material production in 1972.[37]

Roche asked Schadt to prepare a proposal to commercialize the company's liquid crystals. They also asked Schadt to develop a program to license the TN-LCD patent to the emerging liquid crystal display industry, and then put their lawyers to work filing patent applications in other countries. Within a one-year period after filing their Swiss application, Roche had filed TN-LCD patent applications in twenty important industrial nations, with the exception of Korea and Taiwan. With their legal and scientific experts unable to persuade Bauer to change his mind on issuance of the U.S. patent for the device, they decided to fight Jim for it instead.

> *"The patent system added the fuel of interest to the fire of genius."*
> —Abraham Lincoln
> (the only U.S. President to hold a patent;
> U.S. Patent Number 6469, Buoying vessels over shoals)

15. HEARTBREAK AND ANXIETY AT ILIXCO: 1973–74

"We invented Valium, so we have deep pockets."
—Hoffmann-La Roche lawyer to Jim Fergason
in court fight over control of the TN-LCD patents

THE SECOND PHASE of the patent battle erupted in the fall of 1973, five months after Ilixco was awarded the TN patent. Hoffmann-La Roche sent a patent attorney named Bernard Leon to meet with Jim. Leon asked for a license to the issued TN-LCD patents. He stated that the company had their own TN-LCD patent application in process, and although they wanted a license, they didn't consider it all that necessary. The patent portfolio was now Jim's intellectual property and he could choose to license it to other companies. However, he had refused prior inquiries from both Merck and Seiko because if these companies held the license, they would have the right to create and market products based on the intellectual content in the patents, in competition with Ilixco. Jim also turned down Bernard Leon's request.

That's when Leon put the pressure on. He said that his company had "deep pockets" (they had made more than $2 billion in US sales Valium and Librium) and that if Jim didn't give them the license, they'd contest the U.S. patent and tie it up in the patent office for years. "We have the money to take this fight to the next generation," he added. Jim bristled at the bullying, said "hell no" to the license, and told Leon to "bring it on."

Meanwhile, Ilixco was meeting their contract for Gruen watches, delivering about 200,000 TN-LCDs displays. They also delivered non-depolarizing reflectors with tiny edge lamps, obtained from a one-man operation located in one of the Carolinas.

In order to manufacture watches, Gruen had ordered watch cases from what was, at the time, their only source in the world: the Swiss. Gruen couldn't finalize any watch products at all while they waited for the cases to arrive. It's possible the Swiss wanted to protect their domestic watch industry and had no intention of delivering the cases because they never arrived. As a result, Gruen realized their watches were not going to be built and decided to return the entire TN-LCD inventory to Ilixco, claiming quality issues with the displays. Jim only later found out that the real reason was that they could not get the cases as well as enough circuits to match the displays.

American watchmakers had been struggling to regain market share since WWII. Gruen was an old, venerated family business but after the family sold their share in the business in 1953, the company declined. In 1958, the company was split and the watch portion of the business sold to new owners, who moved the operation from Cincinnati, Ohio to New York but kept the Gruen name. The watchmaker was in financial trouble. All of these factors played into their decision to return the displays.

The returned inventory arrived on the day before Thanksgiving of 1973. Ilixco had decided to give each employee a turkey as a holiday gift. The frozen turkeys had just been delivered and piled up on the shipping dock when all of the TN-LCDs sent back by Gruen arrived. They were stacked up on the receiving dock next to and on top of the turkeys, accompanied by a report stating that the displays were 100 percent defective. In other words, they were turkeys.

The staff didn't know whether to laugh or cry, but it soon became clear they should cry. The return of hundreds of thousands of dollars' worth of product ended Ilixco's relationship with its largest customer. Ilixco had to absorb this loss. The substantial amount of money borrowed from National City Bank of Cleveland to finance the expansion, equipment upgrade, and hiring to fulfill the Gruen order had left Ilixco with no reserves. They lost their security interest at the bank and were left living on margin. Gruen went out of business two years later.

The stress got even worse when it appeared that Roche was going to make good on their threat and contest the U.S. patent for the twisted nematic. In an attempt to avert a legal fight, Jim travelled with his lawyer Tom Murray to Europe in early 1974 to meet with Roche's attorneys. However, the meeting quickly spiraled into heated argument and ended without a resolution.

Desperate for cash, Jim and the board made the bitter decision to sell their patent portfolio to Hoffmann-La Roche for the royalties. At first they considered selling only a portion of their TN-LCD-related patents and holding the "gateway" applications in reserve. Roche was not aware these gateway applications existed because the patents had not yet been issued. Once a patent issues, it is in the public domain. When the gateway patents issued, they would dominate the licensed patents, allowing control of the technology to remain with Ilixco. The board considered this a viable strategy because they felt Roche's legal team was technologically unsophisticated and did not have adequate support from technical personnel. They might never pick up on the fact that the gateway patents were not yet available.

However, in the end, because of American tax law, Ilixco decided to sell all of their patents. Funds from an outright sale would be taxed as capital gains at a much lower tax rate than funds from the sale of a license. The sale was negotiated in May 1974 for $1 million, to be split into two $500,000 payments. A trust was formed to hold the funds. Roche made the first payment immediately. They also promised to pay half of all U.S. royalties and a smaller percentage of foreign sales.

The deal included five other provisions that would benefit Ilixco to a large extent. First, licensees in the United States were to pay a fee of roughly 2 percent on sales in the U.S. Second, of those U.S. generated license fees, the Trust would get half and Hoffmann-La Roche half, after the expenses of paying for an agent to collect the license fees were subtracted. Third,

Ilixco got less of a share of the license fees earned in the rest of the world. Fourth, Jim negotiated a grant-back license for the TN-LCD patent portfolio that would allow Ilixco to continue making displays without having to pay royalties. The final provision was that the liquid crystal materials he and Sardari had developed remained the company's property. This was significant, in that Jim later used improved esters of the type pioneered by Sardari to develop surface mode and encapsulated nematic films. Sardari was a co-inventor with Jim on the liquid crystal ester patents, and inexplicably assigned his rights to Roche. Sardari's other partners couldn't understand the legal basis of his decision, but as a chemical giant, Roche was on friendly terms with Kent State University and, with Ilixco failing, Sardari may have hoped that KSU would hire him back for bringing them this portion of the deal.

Ironically, a month after Jim sold the patent portfolio, Dr. Hamamoto of Seiko contacted him to discuss a deal to buy the patents. The board would have preferred selling the patents to friendly Seiko rather than hostile Hoffmann-La Roche, but it was too late.

KSU MAKES CLAIM OF OWNERSHIP

That summer, KSU provost Bernie Hall visited Ilixco with a patent attorney from Cleveland named Lowell Heinke. The pair said they had heard about the company's success and were interested in what they were doing. Jim was too busy to talk to the men and asked Tom Harsch to show them around the plant. Tom was proud of the company's accomplishments and enjoyed showing visitors their research and manufacturing operations. He made it a point to introduce the visitors to former KSU employees or alumni, including Sardari Arora, Tom Davison, Karen MacDonald, and Duane Werth.

After the tour, Hall and Heinke joined Tom in his office, where Tom described the technical progress that had been made. The men showed genuine interest and talked to Tom for an hour. When they were ready to leave, Heinke excitedly said to Hall that he thought that what they had heard confirmed that the TN-LCD came out of the Liquid Crystal Institute. That's when Tom realized their visit wasn't as innocent as they had made it seem.

Soon after, Kent requested that an Ilixco representative attend a meeting in the office of Kent's dean of research, Eugene Wenninger. When Tom Murray and Tom Harsch arrived for the meeting, they saw that Lowell Heinke was there, too. This time Heinke was introduced as an assistant attorney general for the state of Ohio working on matters related to licensing university inventions and copyrights. Heinke announced that he believed the university had an ownership interest in the TN-LCD patent and that he planned to take Ilixco to court over this.

THE BACK STORY TO KSU'S CLAIM

There is a complex back story to this claim. When Jim had made improvements to dynamic scattering material for the Timex contract with LCI, he sent a memo to Hall and Glenn Brown stating that the materials he had developed for this contract were potentially patentable. Correspondence from the LCI files between Daniel Jones and Bernard Hall, dating from 1971, shows that Daniel Jones then orchestrated a patent agreement between KSU and Timex to enable both parties to profit from any patents that might result from this work. Correspondence and memos from the records show that KSU kept asking Jim to provide extensive information on the work he had done so that Timex could write patent applications, and KSU could share in the profits. Alan Coogan, who was the associate dean of the Graduate School and Research, wrote Jim asking him to describe procedures for preparing new materials to increase the lifetime of dynamic scattering displays, new techniques for aligning liquid crystals, and new additives for improving the dynamic scattering operation of liquid crystals.

Jim must have given them information, as there is a document in the LCI archives containing five Timex patent abstracts (which are done prior to an application).[38] However, whatever information he provided apparently did not satisfy either the university or Timex, and so no patent applications had been made. Another complication was that Timex had never paid KSU for the work Jim and his group had done under contract for them at LCI. KSU asked Timex to make the payment, but Timex refused, claiming LCI had not fulfilled the contract. KSU wanted Jim to provide them with enough details on the work he did to enable the university to get paid. Coogan wrote Vice President Hall that "only Fergason has the capability and the materials (which he took with him) to satisfy Timex."

By this time, Jim had invented the TN-LCD, and KSU of course knew this. Their continued insistence that he provide them with patent disclosures for Timex was possibly a ploy to get Jim to "confess" the invention had been made at Kent State. Essentially, they were claiming that Jim had invented the TN-LCD under the Timex contract. At the least, they wanted to elbow in on his valuable invention.

However, the work Jim had done for Timex was on dynamic scattering materials. Jim wouldn't confess to something he hadn't done. This infuriated Kent State's administrators. They sent their special counsel, O.J. Schneider, to badger Jim in person. In their memos to each other, they began discussing not only filing a lawsuit against Jim, but forcing Ilixco out of business. The last paragraph of a memo from Alan Coogan to Bernard Hall reads, "I should note that Mr. F. feels that he "did this on his own time" but doubt that this view would hold water. In addition, we probably have a pretty big stick [in] the form of patent rights and probably could nearly close down his business, or a goodly part of it."

This was the backdrop to the first meeting the university had with Ilixco. A month later, Jim and Ilixco's lawyer Tom Murray met with Heinke and Wenninger again. Murray asked them if they would be willing to discuss a compromise that would allow Ilixco to stay in business. He appealed to them on humanitarian grounds. If Kent State carried out its threat to take the company to court, 220 employees would likely lose their jobs.

This appeal was in vain. The pair said that a higher principle was at stake and the university would proceed with its intended action. Jim seldom said a bad word about anyone, but after the meeting he remarked to Tom Harsch, "I believe that attorney is a truly evil man." Shortly after, the Ohio Department of Taxation began auditing Ilixco for a state sales tax investigation. Jim suspected, because of the timing, that KSU had instigated the audit.

Ilixco later found out that there was even more going on behind the scenes, involving Daniel Jones, the assistant dean of research at Kent State and the attorney that had, back in 1970, failed to file the paperwork required to incorporate Tensor Liquids. Jones was technology unsophisticated but a real wheeler-dealer. He had negotiated a contract with the Timex Corporation enabling Timex to buy a license for $15,000 from Kent State to all horological applications of the university's liquid crystal technology. Although $15,000 was not a lot of money for a large corporation even in the 1970s, Timex never paid the license fee.

Jones put Ilixco under greater siege by writing an affidavit stating that he had visited Fergason at LCI in early 1970 and that Jim had showed him a TN-LCD and explained the device. It was the same technology, Jones said, that appeared in the TN-LCD patent. Jones proceeded to inform Timex that Fergason had developed the TN-LCD while at LCI and that it belonged to the university. Consequently, by virtue of their license with Kent State, Timex owned the rights to the invention.

Later, Jim found out what had motivated Jones to tell Timex they owned the patent. Jones had cultivated a personal and professional relationship with a Timex senior executive, Norman Zatsky. Jones must have hoped to reap a business-related benefit from this relationship. The idea that Timex owned rights to the invention was ludicrous.

KSU's claim that Jim stole their intellectual property was ironic given that university officials had driven Jim out of his job. If KSU had kept him on, he would have been at LCI when he conceived of the TN-LCD invention and would have published the results of that discovery under their auspices. However, it is also doubtful whether the university would have even applied for a patent for Jim's invention, given that KSU had never patented anything since its founding in 1910. In addition, LCI was not in the business of supporting applied research such as Jim was doing.

Before the invention of the TN-LCD, there were no financial or legal issues between Ilixco and KSU or between Ilixco and Timex. Now, Timex was claiming they had rights to the invention because of their contract for research in dynamic scattering displays with the university, while KSU took the position that they owned the device because Jim had supposedly invented it while he was their employee.

KSU wanted ownership of the TN-LCD patents, and Ilixco was in the way. Eugene Wenninger wrote to the provost that Ilixco was in financial trouble, and suggested the university drive them out of business. The university did just that by filing a suit in the Ravenna, Ohio District Court in August, 1974, asking that the company be liquidated in a Chapter 7 bankruptcy. KSU's lawyers told Jim that KSU owned a piece of the TN-LCD patent and had to put the company into bankruptcy to protect the university's interests. This action destroyed Jim's dreams of building a display industry in northeastern Ohio.

Roche later refused to make their second $500,000 payment to the Trust, due in September 1974, stating that they didn't know whether Ilixco owned the TN-LCD patents or whether KSU and Timex did. They claimed that the patent was invalid and nobody held the rights to it.[39] On the day Jim found out that Roche was withholding their second payment, he still fulfilled a promise he had made to his children to take them to the circus.

A TRAIL OF LAWSUITS

Jim and his partners were angry. They had worked tirelessly on the TN-LCD and weren't about to give it up without a fight. They made the decision to sue Roche for breach of contract. They also decided to sue KSU, Daniel Jones, and Timex for interfering with their contract with Roche. This was a gutsy—some might say foolhardy—decision for a company that couldn't afford to pay rent on the building, let alone afford a lawyer. They began looking for a lawyer who would take the case on contingency.

In response to KSU's suit pushing for a Chapter 7 bankruptcy, Ilixco filed for Chapter 11 bankruptcy in federal court. A Chapter 7 bankruptcy would mandate that the assets of the company be liquidated, which would give KSU ownership of the TN-LCD patent portfolio. It would have enabled KSU to grab all the company's assets, whereas a Chapter 11 bankruptcy would buy Ilixco time, allowing them to keep their assets, retain patent ownership, and bring in income to pay off their debts.

With the cooperation of their creditors, the existing management of Ilixco was left in charge. Almost everyone else was laid off. Because the equipment used to make displays had been pledged as security for the bank loan, National City Bank owned all of the physical assets and had to arrange for their liquidation. In a smart move, the bank offered Jim a 10 percent commission to sell off the equipment. Jim was as broke as the company and accepted the offer.

Liquid Crystal Displays (LXD), a start-up TN-LCD company financed by Dickey-John, bought some of the liquid crystal display manufacturing equipment and set up a display manufacturing operation at the other end of the Miles Road building. The LC materials, alignment coatings, adhesives, and buffing machines stayed with Ilixco. Under Jim's direction, the now skeleton staff made LC material for LXD on a lab-scale, packaged the material in Teflon bottles, and walked the bottles over to them. LXD also hired a number of former employees, and bought the rights to the Ilixco name, although they never used it. The company is still in business. On its home page, the company pays tribute to its Ilixco heritage, writing that "Our roots go back to the International Liquid Crystal Company—the inventors of TN liquid crystal displays."

Ilixco stored the company's remaining physical assets in a local warehouse and vacated the Miles Road building. Jim sold most of the liquid crystal display fabrication equipment to the Hamlin Corporation, a division of Standish Industries. Hamlin was already trying to be in the liquid crystal display business in that the company had been trying with little success to manufacture dynamic scattering displays. With the failure of Jim's business, Hamlin had no competition. They implemented a deliberate strategy to launch the successful manufacturing of TN-LCDs, using Ilixco's equipment and Jim's expertise. They engaged Jim to help set up the TN-LCD line at the Hamlin facility in Lake Mills, Wisconsin, and hired some of Ilixco's employees. Duane Werth interviewed for a sales manager position. His wife, Karen McDonald, who knew the entire display-building process and would have been valuable in helping Hamlin learn it, also interviewed with Hamlin. It wasn't until later that Karen and Duane found out why Hamlin didn't hire her. Duane wrote to Tom Harsch in 2012:

> Hamlin had been working on making LCDs with little success until they started reading the process books and writings on the fiberglass "lunch trays" we [Ilixco] used to transport the raw materials. Apparently one of the production engineers wrote the process on the trays so the production people could follow it. All Hamlin needed to do was follow the instructions on the trays to end up with the entire production process. Within a year they were building displays exactly like ours. It would be several years before they could get the liquid crystal materials from outside suppliers to build a reliable display. In one of the file cabinets were the actual process paperwork files with temperature, times, materials, part numbers, everything needed to duplicate our process. The tray writings filled in some needed production details that were not in the process papers. I think Hamlin paid less than $2,000 for a process that cost us $4 to $5 million dollars to develop. They were a smug bunch and didn't hesitate to tell the story.

The prominent liquid crystal engineers Mary Tilton and Bill Tonar later joined the Hamlin group. Hamlin was eventually sold to Standish (later Sterling), and the liquid crystal division of Sterling was later sold to Planar Systems. Planar stopped making displays in 1998.

During the equipment sale, Jim brought Tom and several other former employees he wanted to work with again to the warehouse to see if there was any equipment they wanted to buy. It took time to find equipment, because it was in boxes spread out over 20,000 square feet. Numerous former employees now working at the LXD startup were also hunting for equipment. Jim was appalled when he observed several former employees pocketing small pieces of equipment. He had always taught his children that stealing was a character flaw. Perhaps, he told Dora later, these former employees justified stealing things they thought would help them perform well on their new job, but he felt betrayed.

Tom Davison, who had been running Liquid Crystal Biosystems, acquired cholesteric liquid crystal thermography know-how, laboratory test equipment and some chemicals in exchange for back pay and unpaid expenses. Davison went to Chicago, where he partnered with Fred Suzuki to co-found Liquid Crystal Products, which later became BioSynergy. Duane and Karen (McDonald) Werth also moved to Chicago, and Karen joined Tom Davison at his new company. Dick Strong and Duane both worked for a liquid crystal display distributor.

Jim was already thinking of starting another liquid crystal-related company, and purchased the last of the equipment, including the chemical synthesis equipment used in the pilot plant. By this point, it was quite a bargain. However, he had little time to devote to starting a new business because he was ready to proceed with the lawsuit against Hoffmann-La Roche, KSU and Timex.

16: THE TWISTED NEMATIC LAWSUIT: NOVEMBER 1974 TO JANUARY 1976

> *"Bringing a new invention to market is a full-contact sport."*
> —Jim Fergason

JIM HAD BEEN looking for a lawyer willing to take the case on contingency. He initially asked John Burlingame, who had served as a non-voting secretary on Ilixco's board and was also Ilixco's general council and a partner with Baker Hostetler, one of Ohio's leading law firms. An aggressive former fighter pilot, he would have been an excellent choice to lead the charge into court, but it was a Baker Hostetler policy not to take any case on contingency.

Realizing that it would be next to impossible to find anyone else, Jim asked his reliable friend, Pittsburgh attorney Tom Murray. Jim's request must have given Murray pause. Right had prevailed over might when the U.S. Patent Office had awarded Jim the U.S. patent for the twisted nematic over the competing application filed by Roche. But the patent review process was an impartial one, and while the court case would be, too, it was one thing to have a decision made by a technical expert at the patent office and another to have one made by jurors who would not necessarily have the technical or scientific background to understand the case. Murray would also be facing off against three daunting opponents with enormous financial resources.

Nonetheless, Murray agreed to take on the case. He recommended that Ilixco hire a second attorney, both because the case would generate an enormous amount of legal work and because it would be tried in Ohio and Murray needed an associate who was licensed for the Ohio bar. However, there was no money to pay a second attorney. The only attorneys Murray could find who were willing to take the case on contingency were Clevelanders Martin Goldberg and his nephew Richard (Rick) Goldberg. The downside, however, was that the Goldbergs were personal injury attorneys with no experience litigating patent cases, but they took on the case.

With three lawyers lined up, Ilixco proceeded to sue Kent State University, Daniel Jones, and the Norwegian entity of Timex for interfering with their contract with Hoffmann-La Roche. They also sued the U.S. Hoffman-La Roche Nutley, based in Nutley, NJ and the Swiss entity of Hoffmann-La Roche (Hoffman-La Roche Basel, in Basel, Switzerland) for breach of contract, racketeering violations under the RICO act and restraint of trade. The total amount of damages sued for was $85.5 million.

In an article on the lawsuit published in the December 4, 1974 issue of *The Sun*, a Cleveland-based newspaper, Jim refuted KSU's claim that he developed the technology while at LCI. He "acknowledged that he developed five other patents for digital watches while working for the institute," and that "these patents are the property of KSU and Timex, Inc." Murray told the newspaper reporter that the patent was extremely valuable and that the company had spent about $4 million developing a market for the invention. KSU's patent lawyer, Lowell Heinke, said that "the university had allowed Fergason to engage in outside work as long as it did not conflict with his work at the Institute," but that "the patent in question is a result of Fergason's work at the institute." Timex and Roche declined to comment for the article.

Murray felt the racketeering case was uncertain because Ilixco was "perceiving smoke and using a trial in hopes they could find the fire," but he had confidence that the restraint of trade case was a strong one. The basis of this case was that Roche had acted in concert with Swiss watch companies to protect the Swiss watch industry by driving Ilixco out of business. As Jim told *The* Sun, the actions of both Roche and Timex were a "conspiracy to drive Ilixco out of business and dominate competition for the Swiss watch industry, and for Timex, largest U.S. manufacturer of watches."

Martin Schadt had worked for the Swiss watch company Omega in 1969, just before joining Roche. Schadt later stated in an interview published by *Advanced Imaging* that at about the time Fergason filed his lawsuit against Roche, the company was considering withdrawing, for business reasons, from liquid crystal-related businesses. It must have seemed ironic to Roche that they were now involved in a potentially expensive and time-consuming lawsuit.

Jim later learned that Roche had been surprised he filed a lawsuit against them because the corporation was the largest pharmaceutical company in the world, and had a reputation for running roughshod over inventors. Ilixco was like David fighting Goliath, and they were doing so with the significant handicaps of inexperienced lawyers and no money to hire expert witnesses. At one point during the trial, Murray and the Goldbergs faced a total of 18 lawyers from the other side as Roche had hired the best legal firepower in the patent litigation field that money could buy.

Murray requested a jury trial. His opponent's lawyers responded by suggesting that the issue of patent ownership be litigated first in a separate case since it was a matter of law, not of equity. As such, the decision on patent ownership should be made by a judge. Only after the issue of ownership was decided would the trial address proceedings related to contractual rights, and then to the RICO and restraint of trade cases. On the surface, this suggestion sounded reasonable and Murray agreed. This could turn out to be a terrible mistake, because there would be no financial settlement if Ilixco won the first case and those legal expenses alone could use up all of the company's available cash, making it impossible to fund the next case on contractual rights. Ilixco would be unable to make it to the next pay day even if the first case went well.

THE TRIAL BEGINS

Nevertheless, Jim took his chances and began what would become a costly, complex legal fight. The case was tried in the U.S. District Court in Cleveland, with Judge William K.

Thomas presiding.[40] A lawyer, Zachary Paris, who clerked for Thomas told us that while liquid crystals have today become a huge industry, at the time of the trial, "we didn't know what liquid crystals were."

However, Judge Thomas had an "insatiable desire to understand how things worked," and was "relentless in his pursuit of the facts underlying the disputes that came before him." As the trial proceeded, Jim got to experience for himself why Judge Thomas had a reputation for integrity. Thomas was "straight as an arrow, and the quintessential jurist," his former law clerk said. *The Plain Dealer* used to take polls of active members of the bar on their opinions of judges, and Judge Thomas always topped the list. "He was the best I've ever tried a case in front of," a veteran Cleveland civil defense attorney said of Thomas. "He was patient, he listened to what you were saying and then ruled decisively. That's the whole package."[41]

The only complaint lawyers had about the judge, as Zachary Paris told us, was that he "refused to let a litigant get a raw deal because of an incompetent lawyer. He would ask questions during trials that an incompetent lawyer didn't think to ask. He knew how not to let a key moment pass in a trial. The rules of federal procedure allowed judges to ask questions, and he asked them." This annoyed experienced lawyers who were facing incompetent ones and wanted to use their experience to advantage in court.

Soon after the trial began, one of Roche's attorneys contacted Ilixco, an unusual move for an attorney on the opposing side. The attorney reported that Rick Goldberg had attempted to solicit a bribe for $100,000 to influence Ilixco and obtain a settlement favorable to Roche. Because Goldberg was on contingency he couldn't be fired, but Ilixco did not let him appear again in court. If they had pursued legal recourse, Goldberg could have been disbarred and possibly found guilty of a criminal act. Years later, he was sent to prison for criminally defrauding clients.

KSU was represented by patent attorney Lowell Heinke, while Roche was represented by a New York-based law firm that their U.S. office routinely hired for legal matters. Their chief council was Bernard Leon, the attorney that had tried to strong-arm Jim into licensing Roche the patent. The U.S.-based Roche employees had interacted with Jim when he was at LCI, and both respected and liked him. They disliked their counterparts in Roche's European headquarters. During the trial, these employees sometimes provided Jim with information about what was going on behind the scenes at Hoffmann-La Roche.

As part of the discovery process, Ilixco allowed the opposing attorneys full access to all company documents, including laboratory notebooks. Roche did not reciprocate. The obstacles Murray faced in obtaining information were compounded by his lack of financial resources. Ilixco couldn't afford to send him to Roche headquarters in Switzerland to review documents.

The lawsuits took 14 months to resolve. By the end of it Jim was physically, emotionally and financially drained. As a key participant in the proceedings, his trial experience was a debilitating blend of tension and boredom. It reached the point where he would have liked to walk away from the entire case. However, he never really seriously considered bailing out. Beyond a powerful sense of knowing he was right, he felt a strong obligation to the company's investors.

MAKING MONEY WHILE THE TRIAL GOES ON

One of the many stresses Jim faced was that he had to provide for his family while shouldering the considerable demands of the trial. Although he had paid into unemployment, he couldn't collect it because he was the principle owner of Ilixco. Judge Thomas called occasional recesses in the trial to give Jim time off to earn a living. Jim made the most of this time by setting up a new liquid crystal company. He rented space and called several former employees offering them work. However, by the spring of 1975, when Jim's daughter Terri was applying to colleges and needed financial aid, Jim was so close to bankruptcy that he didn't know what to report as his income on her financial aid form.

Dora began working as a finisher for needlepoint stores in the summer of 1975 to help support the family. On the 1970 Europe trip she had admired the huge tapestries she saw in museums, and brought a large tapestry home from Europe to work on. She became as expert in needlepoint as she was in sewing. She also subcontracted with other women to do finishing work for her. The family didn't have a penny extra to spend, and they all felt the strain. The one bright spot was that Dora became pregnant. The family's three teenaged children were as excited about the pregnancy as their parents, particularly Terri, who badly wanted a sister.

DEPOSITIONS AND BACK TO COURT

During the first year of the trial, Jim made depositions to Kent State University, Hoffmann-La Roche Basel, Hoffmann-La

Roche USA, and Timex. A deposition is the out-of-court testimony of witnesses for gathering information. It can consist of written or oral questions. Roche's agenda during the depositions was to destroy Ilixco's patent ownership position by finding a smoking gun, so they questioned Jim intensely. Some of the depositions lasted a grueling four days. Between the discovery and the depositions, the case accumulated over 1,700 exhibits, including books and sales records. "We're going to bury you in paper," one of Roche's lawyers said to Jim during the trial, and bury him they did. Piles of documents a foot high covered three work tables. By the time the case ended, the case records had to be stored in five boxes, each 5 cubic feet.

The in-courtroom portion of the legal proceedings lasted four weeks, with regular but short interruptions to allow the judge to handle higher priority matters involving criminal cases. Jim testified at the trial and was on the stand for two weeks. This was even more exhausting than giving the depositions.

At the beginning of the trial, Roche's lawyers tried to show that Jim had invented the TN-LCD at LCI just after the Kent State shootings. Judge Thomas had presided over several different civil trials concerning the Kent State shootings, the first in 1971. He knew that Jim could not have been on campus because the university had shut down after the shootings and barred its staff, faculty and students from campus. Then and there, Thomas resolved the inventorship issue in Jim's favor.

After Thomas declared Jim the inventor, the trial entered the next phase focused on the ownership of the TN-LCD patent. Glenn Brown, KSU president Robert White, and Daniel

Jones all testified on behalf of Kent State and Hoffmann-La Roche. Even though Sardari was still a part owner of Ilixco, he testified on the witness stand that Jim did not invent the TN-LC device there. He claimed he saw the device while he and Jim were still working for LCI.

There were political reasons for Sardari's decision to testify for the opposing side. Kent State likely promised him a teaching job at the university if he agreed to testify for them, and he was given a job at a branch campus. But his testimony was of no help to the parties opposing Jim and destroyed Jim's trust in Sardari. From then on, Jim viewed the chemist as a traitor. Sardari claimed that he had worked on Ilixco matters while at LCI, but Jim had specifically instructed Sardari not to do so, and he almost certainly complied. Tom Murray elicited testimony from Sardari that Peter Pick of Hoffmann-La Roche had synthesized liquid crystal materials for Tensor Liquids in his garage, implying that Ilixco did not need to use the facilities at LCI for synthesizing liquid crystals. Under cross-examination by Tom Murray, Sardari became confused about the optics and physics of liquid crystal displays and was discredited as a witness.

Roche called Jim's former boss, Glenn Brown, to support Roche's position that the TN-LCD was invented at LCI. However, Glenn also proved a bad witness. He said that Jim was not involved in applied research but only in basic and military research. He referred to a written definition of applied research and to piles of memos he sent to Jim on the subject. He further stated that he had banned applied research at LCI. Given Judge Thomas' "profound respect for the law," it was likely that he knew of the 1933 case, U.S. vs. Dubilier Condenser

Corporation, in which it was established that an employer only had rights to an invention if the employee had been "specifically hired to invent" as opposed to hired to do research.

A parade of witnesses from Kent State continued. KSU president Robert White testified, but he didn't know any specifics and had nothing to contribute. Daniel Jones' testimony consisted of what he had written in his affidavit, that Fergason showed him a TN-LCD at LCI in early 1970. On the stand he proved inept, becoming confused about what he had seen during his visit to LCI, the technical content of his discussion with Jim and the subject matter contained in the TN-LCD patent. In addition, his testimony seemed to violate attorney-client confidentiality, because Jim had hired and paid Jones to incorporate Tensor Liquids, although Jones never followed through with this. This put Jones in legal jeopardy, but Ilixco waited too long to file a formal complaint and it never became a legal matter.

While researching Kent State's involvement in the lawsuit, the public services librarian at Kent State looked through the LCI archives as well as the personnel files of the individuals involved for any mention of the patent case. The librarian found no documentation whatsoever of the lawsuit in the archives, and remarked that she and the other librarians "often wonder if some materials never made it to the archives because individuals thought they might be too inflammatory." There is an extensive paper trail in the archives about Jim Fergason and Ilixco, but it is all dated before and after the lawsuit.

Ilixco's lawyer Tom Murray called to the stand Alfred Saupe, Dwayne Baumgartner, Jim Bell and Tom Harsch to testify. Alfred confirmed that, to his knowledge, development activities directly related to the TN-LCD were underway at

Ilixco at the time claimed by the company. Dwayne Baumgartner, the CEO of Donnelly, testified on Ilixco's contract with his company. He specifically stated that the TN-LCD development program was funded by Donnelley. This statement was directed against ownership claims by KSU. Baumgartner also stated that he personally saw an operating TN-LCD prototype at Ilixco in the fall of 1970, before Roche filed their Swiss TN-LCD patent application. Along with other evidence, this proved that KSU had no claim to ownership.

Although nervous in the unfamiliar courtroom setting, Tom Harsch proved a star witness, providing specific firsthand information on the events and timeline related to the development of the TN-LCD. Investor Jim Bell also testified, but mostly on financial matters. He later recalled that he presented his testimony without any preparation from Ilixco's attorneys.

Notably, Roche's employees who had claimed to own the patent, Martin Schadt and Wolfgang Helfrich, did not testify during the trial. Helfrich provided a statement in which he "swore back" his invention of the TN-LCD to late 1969 or early 1970, when he was working at RCA. Roche wanted Jim to take a lie detector test, implying that they felt Jim was lying about when he invented the TN-LCD. Jim said, "I'll agree to take it if Schadt and Helfrich do." Roche dropped the subject.

THE JUDGE WORKS TO SETTLE THE CASE

At this point in the trial, Roche began negotiating with Timex and KSU to buy out their positions in the case. The cost of the case was certainly a factor. The Ohio attorney general told KSU to stop the litigation because it was too costly, while

Timex had already spent an unaffordable $1 million in legal fees, putting them on shaky financial ground.

Ironically, one complication prevented the deal from proceeding: Ilixco was suing Timex. Timex made Ilixco an offer of $100,000 to be allowed out of the lawsuit. Ilixco accepted, and agreed to settle the case only with Roche. During the negotiations, the Timex lawyer made an unusual off-the-record comment to Jim that Jim was the only party in the whole legal mess that had contributed anything. Jim got the impression that the lawyer really did not want to be a part of the case. As part of the agreement, Timex was also required to pay the outstanding invoices of $14,000 that they owed KSU for the research Jim had done on dynamic scattering under their contract with LCI.

Judge Thomas helped negotiate the buyout agreements. He also encouraged the two remaining parties to reach "an amicable settlement among all of the litigants of the differences between them." A master mediator, he placed the legal teams from the opposing parties in separate rooms at the courthouse, and went back and forth between them. "This was the essence of shuttle diplomacy," Zachary Paris told us. "He did this so each side could speak more frankly to the judge, and the judge could speak honestly about the strengths and weaknesses of their position. He had a lot of credibility with the lawyers."

Judge Thomas told both parties that litigation of the existing suits would take a long time and be very expensive. He told Jim that he suspected Roche would appeal if they lost the case, doubling the time and expense associated with it. Finally, he urged Jim to get back to inventing because all the legal "wrangling" had diverted Jim's attention from research and development activities for well over a year. This was not only a loss to Jim personally, but to the scientific community.

Between pressure from the judge, the possibility of continued litigation at substantial expense and the fact that they were losing the case, there came a point where Roche offered to settle. Jim was also ready to settle. He was on the verge of personal bankruptcy and could not devote more years of his life and career to litigation. He had learned that patenting something and controlling the patent were very difficult.

After all the trouble and expense, the proposal brokered by the judge was that, essentially, Ilixco and Roche were to revert to their original agreement. This meant that Ilixco's ownership of the TN-LCD patents would be recognized, Roche would meet their contractual obligations to Ilixco, who would drop the RICO and restraint of trade lawsuits. There were three additional provisions.

1. KSU was to receive 10 percent of the royalties.
2. Timex was to get a limited, prepaid royalty for the TN-LCD.
3. Ilixco was given a license to what was now Roche's TN-LCD patent portfolio, and had certain rights to sub-license the technology.

Under these terms, there was little incentive for Jim to agree to the proposed settlement, as it would leave him still broke and in possession of a bankrupt company. The only way he could come out ahead financially was if the RICO and restraint of trade suits were to proceed and Ilixco to win a cash settlement, because then money from the settlement/penalty would be paid to the Trust.

As part of the distribution of funds to stockholders, he would receive some of this money.

Jim brought this problem up and it was quickly resolved by his own lawyers. Like all other parties involved in the lawsuit, the trial had financially and emotionally exhausted his lawyers. They offered Jim $25,000 out of their own pockets to agree to the settlement. This was a practical offer rather than a good will gesture. The lawyers' ability to recover financially depended on reaching a successful settlement and being paid their contingency fees.

BACK AT HOME, SUSAN MICHELLE IS BORN

Throughout the trial, Jim had commuted between his home in Kent and the courthouse in Cleveland. In December of 1975, Dora was entering her eighth month of pregnancy. She worried that Jim would be in Cleveland when she went into labor, especially since all of her children had been born early. But the baby was born on December 23, 1975, when Jim was home for Christmas. Terri got her wish for a sister, and the family their best Christmas present ever. They'd had no life beyond that of the case, and the baby's arrival gave them joy. The Fergasons allowed their three older children to choose the baby's name. After some discussion, the children named her Susan Michelle.

THE CASE IS SETTLED, THOUGH JIM REMAINED UNCREDITED

A week after Susan's birth, Ilixco and Roche solidified the agreement over the terms of the final settlement. The lawsuit was settled by court order over the judge's signature. As part of the settlement, the trial records were sealed and all parties were prohibited from speaking of confidential matters in public. After the patent trial was over and the payments made, Ilixco's stockholders were given the option of accepting shares in the trust Ilixco had formed to hold its assets. The funds were distributed in January 1976.

It's impossible to know how Judge Thomas would have ruled if the case had proceeded, but Jim thought he would have ruled for Ilixco. The lawyers already had a big win, because the legal fees for all of the participants exceeded $5 million.

After the trial, Roche publicly claimed victory. In Northern Ohio, the home of Ilixco and KSU, almost nothing was reported, in part because of a newspaper strike. The KSU student newspaper, *The Daily Kent Stater*, published a short article on the lawsuit on January 13, 1976. The article erroneously reported that KSU had accepted a $100,000 out of court settlement with Roche "in exchange for any rights to liquid crystal watch display devices," and that the "settlement provides for the university to receive a percentage of future royalties from the company." However, the article failed to report Judge Thomas' crucial finding that KSU had no claim to the invention. The article instead repeated KSU's assertion that Jim had developed the TN display while working for LCI, implying that Jim took something that didn't belong to him. Jim found this particularly galling.

The February 2, 1976 edition of *Electronic News* published an article reporting resolution of the lawsuit, but it contained even more inaccuracies than the *Daily Kent Stater* article. There was very little interest in the case or its disposition. This was surprising given the importance of the technological property involved. The Ilixco founders were disappointed, feeling that if there had been coverage, it would have been supportive of their position.

The skimpy coverage could certainly be one reason why, to this day, Helfrich and Schadt are credited in numerous print and online sources with having invented the twisted nematic liquid crystal display. In some of these sources, Jim is credited with "contributing greatly to the invention," but other sources either don't mention him at all, or present a revisionist history. In addition, Roche had the public relations manpower to influence what was written about the history of the invention. As Jim's colleague Tom Harsch put it, "history favors people who have the backing of a major institution, which often has a longer lifetime than a human being."

In 1983, a physicist who worked for Hoffmann-La Roche wrote in a letter published in *Physics Today* that, "Fergason obtained a U.S. patent despite the earlier priority by Schadt and Helfrich because the U.S. patent law gives preference to U.S. inventors."[42] This was an inaccurate statement, as anyone familiar with U.S. patent law will attest to. Roche's application was rejected in the U.S. because it was determined to be a restatement of the contents of the *Electro Technology* article. It was later rejected on the same basis by the patent offices in Germany and the Netherlands. In the 2004 book *Crystals That Flow: Classic Papers from the History of Liquid Crystals*, the authors wrote "Fergason subsequently filed a patent in the United States in April 1971 which describes the twisted nematic display, but Hoffmann-La Roche had the edge and bought Fergason's patent rights."[43] Roche had purchased the Ilixco portfolio because they had a financial edge, but certainly not a legal one.

Jim was particularly incensed by the account of the invention in a 2005 article written by Gerhard Buntz, a European patent attorney. The article gives Hoffmann-La Roche's party line version that Helfrich and Schadt invented the TN-LCD in "about September 1970." Buntz wrote that a "liquid crystal researcher from Kent State University" (clearly referring to Alfred Saupe) had visited the BBC labs in Europe where he was told about Helfrich and Schadt's invention. When Alfred returned to the U.S., he told Jim about the invention, and Jim then started making displays. Buntz essentially claimed that Jim stole the idea. He doesn't mention in his analysis that he had worked as a patent attorney for F. Hoffmann-La Roche AG in Basel.[44]

To Jim's dismay, Kent State University persisted in asserting that Jim invented the TN-LCD while he was associate director of LCI. This myth is widely perpetuated by other sources. Even a Smithsonian Museum history of the quartz watch begins, "James Fergason, while Associate Director of the Liquid Crystal Institute at Kent State University in Ohio, discovered the twisted nematic field effect of liquid crystals in 1969." Jim took some solace in the knowledge that he wasn't the only inventor with a university affiliation who had gone through a nasty legal fight over potentially lucrative intellectual property.

Kent State University still calls LCI the "birthplace of liquid crystal displays," although Jim's name is nowhere to be found on the LCI website. In the section on the website about the history of LCI, someone wrote that in the early years of LCI, "Major grants came from the National Institutes of Health, the National Science Foundation, and U.S. defense agencies." There is no mention that all of these grants were awarded to James Fergason.

HOFFMANN-LA ROCHE PAYS LICENSE FEES...FOR A WHILE

After the lawsuit ended, Roche entered the liquid crystal material business. However, the company had a strategy of providing favorable financial terms to licensees in conjunction with other business arrangements between the companies involved. This strategy increased Roche's profits, but effectively decreased the royalties they had to pay to the Ilixco Trust. According to the terms of the legal settlement, Ilixco had no control over Hoffmann-La Roche's TN-LCD patent licensing program, and, as much as they wanted to, couldn't protest this strategy.

Over the next few years, Hoffmann-La Roche paid just $4 million to the Trust. Out of this amount, the people and organizations that had accepted stock in the Trust as payment for money owed to them during Ilixco's bankruptcy proceedings were reimbursed.

Kent State received part of the settlement money, but did not use it to provide funding for LCI. It was only after Glenn Brown retired in 1983 and Bill Doane became the Institute's director that some of the settlement funds were used to support the Institute.

In 1992, Roche missed a royalty payment and when Ilixco's attorney called their attorney to ask why, he was told that

Roche felt they had "paid enough." The lawsuit had worn out Jim and all other parties on the Ilixco side, and they decided not to go after Roche. Roche never made another payment to the Trust and it was closed in 1994.

When Ilixco's patents expired, Roche decided to stop selling liquid crystal materials. Wolfgang Helfrich left Roche soon after the case was settled. There was speculation that he had been fired, possibly because management believed he had acted improperly in claiming invention of the TN-LCD. On more than one occasion, he remarked in public that his departure from Roche related to his "talking too much." Roche treated Martin Schadt dramatically different from Helfrich, however. They promoted him in recognition of his contribution. He also received a 25,000 Swiss franc bonus for the development of the TN-LCD.

All told, Jim never made much money on the TN-LCD. By the time he received the royalty money from Roche and paid his lawyers, he made only $1 million off what is now a $10 billion a year industry.

17. STARTING THE AMERICAN LIQUID XTAL CHEMICAL CORPORATION

"Jim was the Thomas Edison of liquid crystals."
—Dr. Beth Cunningham

WITH THE ORDEAL of the lawsuit finally over, Jim was able to devote all his time to the new company he had formed from the ashes of the old. He drew inspiration from remembering how his paternal grandfather, after gambling his farm away in Indiana, started over in Missouri and became a financially successful farmer. "I identified with his plight," Jim said. "At a time in life when he should have been settled in, he had to start over. He made the best of a bad situation and came back a stronger person." Jim vowed to do the same.

He called the company the American Liquid Xtal Chemical Corporation (ALX). His business plan was to be a liquid crystal materials and technology provider and consulting firm for both U.S. and international display production companies. There was a thriving liquid crystal display production industry in the

U.S. in the late 1970s and 80s, due in no small part to Ilixco's pioneering work in display technology. The companies making displays included LXD, the company that had purchased the display manufacturing equipment as well as the rights to the Ilixco name, Crystaloid, Gor-Vue, and Hamlin. These companies either bought liquid crystal materials from Roche at a premium or made their own. Jim wanted to provide lower cost liquid crystal materials, gasket adhesive alignment layer material and even buffing machines to buff glass with PVA.

He had three other goals: to promote TN-LCD technology and enable its wide adoption, to undertake activities that would maximize TN-LCD royalty payments to the Ilixco Trust, and to provide consulting to companies developing new liquid crystal-related technologies and products.

Tom Murray financed the new start-up, and Jim provided the technology by licensing his own patent portfolio. His decision to license rather than transfer ownership of the patents was a result of his hard-won lessons from the past. He would retain ownership of his future patents regardless of ALX's fate.

During one of the breaks Judge Thomas had called during the trial, Jim had rented an A-frame house on Main Street in Kent for ALX to use as an office. Tom Harsch and Ilixco alumni Hugh Mailer, a Scottish national and PhD. in physics, were among the first employees. They worked for free for six months to help get ALX going, living on the unemployment they collected after Ilixco closed.

Jim offered other former Ilixco employees jobs in February 1975. He hired Kenneth Marshall to work as a chemist, and Howard Sanders, an un-degreed chemist, to work as

Ken's technician. Jim initially wanted to put the lab in a more fireproof building than the A-frame and found an old service station garage in Mantua that he thought might work, but Ken took one look at the interior and knew that without renovation, they'd never be able to set up a lab in this space.

They ended up storing the equipment Jim had bought from Ilixco in this garage and setting up the lab in the basement of the A-frame. Ken and Howard put a microscope with a hot stage for measuring transition temperatures in the finished main room of the basement and a workbench and a fume hood with ducts running out of the window in the other basement room. Don Hurt, a former production manager at Ilixco, made buffing machines in his own shop for ALX to sell.

Jim's children pitched in, too. The first summer ALX was in business, Terri washed bottles for the chemists and answered the telephones to earn money for college, although that fall when she started, her father was still so strapped that he couldn't pay her tuition bill on time and the college only let her register for classes on an IOU. The following summer she was promoted to lab tech. Her brother Jeff did everything from sweeping floors to moving equipment to mixing chemicals. Jim's youngest son, John, had an interest in electronics and worked with Tom Harsch.

ALX was along the route of the neighborhood's Jehovah's Witnesses. Howard Sanders, who was an atheist, always ran to the door when they knocked to argue with them about religion. Howard soon tired of commuting the nearly 40 miles from his home in Cleveland to Kent and so he quit. His replacement, Sharen Breyer, was a smart young chemist who always spoke her mind. Whereas her predecessor liked to argue with the Jehovah's Witnesses, Sharen liked to argue with Jim. She marched around the lab with her hands on her hips. The other techs always knew which lab coat was hers by the smudges on the sides where her hands rested.

MAKING MONEY FROM LC'S — ANYTHING LEGAL WOULD WORK

Sharen helped Ken make compounds and cholesteric liquid crystals and mix up adhesives for displays. "Anything legal we could do to make money, we did it," Ken Marshall told us about those early days. "We sold circular polarizers to the University of Rochester for a laser, and even made and cleaned mood rings." Mood rings, which were then made of small pieces of plastic or glass with the liquid crystal sandwiched between them, became a huge fad in the 1970s. ALX made thousands of them. Jim bought rubber molds to hold the clear plastic mood ring blanks and hired several people to run a manual production line. The chemists mixed the cholesteric liquid crystal material and the workers painted it on the backs of the stones and then sprayed them with black spray paint. None of the workers wore a mood ring themselves. They were sick of looking at them, and mood rings and solvents didn't mix.

Other customers who wanted to make money off the showy properties of cholesteric LCs in response to temperature sought out ALX. A hard-rock band named "The Fluids" bought liquid crystal body paint to slather on for their shows.

A New Yorker named Bob March, hoping to capitalize on the mood ring craze, contacted ALX about making "mood" clothing. There was some precedence for the idea. Glenn

Brown told a *Cleveland Plain Dealer* reporter in 1974 that, for the purposes of detecting breast tumors, Oregon State University was developing a "brassiere that has liquid crystals built into the texture. The patient puts it on, and the color pattern will read out on the brassiere,"[45] though it's unlikely March knew about this idea when he developed his own. Nobody at ALX thought the technical problems with making mood clothing were surmountable, but March was persistent, and ALX finally agreed to experiment. They tried different mixtures and techniques on several kinds of fabric, but the color changes were muted and the fabric became very stiff, greasy, or both. The material also couldn't be washed or dry cleaned. March eventually gave up on the idea of making mood clothing.

JIM HAS A HEART ATTACK

At the end of April 1976, the stress of the lawsuit caught up with Jim. After work one evening he was spreading fertilizer in the yard and the spreader kept jamming up. He complained that his arm hurt and went into the house to rest. The next morning, he awoke with chest pains. Dora roused John, who was then 14, handed over 4-month-old Susie and sped Jim to Robinson Memorial Hospital. The ER doctor admitted him to the ICU with a heart attack and the remark, "You made it to the hospital just in time." In the mid-1970s there were no stents in use in heart surgery and the damage wasn't repairable. There are three major coronary arteries in the heart. Jim now had only one that was intact.

The close call was frightening. Terri, who was a freshman at Hiram College, got a ride home so she could help care for

Susie while her mother spent more time at Jim's bedside. He was in the hospital 10 days. The doctors kept him on Valium the entire time because they wanted him to remain calm, but he hated taking a drug that Roche profited from. Tom and Jacky Harsch visited him in the hospital, and said he was too "drunk on Valium" to have much of a conversation.

When he was finally released, Dora drove him home. He kept asking her to drive more slowly and she couldn't understand why. "Once I get there, I won't be allowed outside again for a while," he explained. He was an intellectually active 42-year-old and, as Susie put it in recalling that time, he was "always tinkering and constantly coming up with new ideas." Now he had to rest.

The bright side was that he could spend time with his family. Jeff and John were teenagers and more interested in hanging out with their friends, but baby Susie loved being cuddled. And Terri, who had resumed classes at Hiram, didn't hesitate to call her father one day when she was studying chemistry with a classmate. They were stumped by a chemistry problem. "I'll call home and ask my dad for help," Terri said. "He'll know how to solve it."

"Girls!" one of her classmates exclaimed. "They always think their dads know everything!" He was amazed when Terri came back from the phone conversation with a solution to the problem.

ALX THRIVES

Under Dora's devoted care, Jim recuperated, but it took months and he became easily fatigued. Nonetheless, when he returned to work, ALX began to take off. The company made an alignment coating called SA-72, a 2 percent solution of high grade Polyvinyl Alcohol (PVA) and deionized water, that they sold for $50 a pint to Hamlin and other companies. It was deposited in a thin film by spin-coating, dip-coating, or wick-coating, dried at 80°C, and buffed to produce alignment for the LC molecules. Some of these companies needed large quantities of SA-72. The coating became a big money maker and took very little work to make.

ALX also wanted to develop a new, improved nematic LC material they could sell to companies making displays. Ilixco's original formulations had been made with either Schiff base material or ester material. The Schiff base material turned an ugly, yellow color in displays—Jim jokingly called the material "Old Yeller" after the dog in the children's book by the same name—and some of Ilixco's customers had complained about this. While the color of the ester formulation was a more aesthetically pleasing white, the formulation crystallized at very cold temperatures, causing black flecks to appear in the display.

Ken tried mixing these two materials together. He mixed together p-ethoxybenzylidine-p-n butylaniline (EBBA) and p-n-pentylphenyl-p-n-pentyloxybenzoate (PPOB), added p-cyanobenzylidine-p-n-butylaniline (CNBBA) and p-n-pentylphenyl-p-n-pentyloxybenzoate (PPOB) to it, and put the mixture in his home freezer for 24 hours to see if the cold would destabilize it. When it didn't, he decided to leave the mixture in the freezer for a week. To his surprise, it remained in its liquid crystalline state. Ken took these results to Jim, and they used this new formulation for the first material they sold. It became ALX's new standard nematic material. They called it the "K" material, and put a number suffix after it (somewhere between 2 and 10), representing the number of tries it took for Ken to achieve a stable mixture composition.

They sampled "K" to Hamlin. Hamlin liked the material, although the operating voltage wasn't high enough. ALX sold the material for $1 or $2 a gram, much lower than other companies, and garnered a lot of business because of this, including from companies in Europe and Taiwan. They also shipped the material to Toshiba in Japan. Later, they discovered that Toshiba analyzed the material, broke it down, and made it themselves, which was illegal.

Jim claimed that the material was superior to the cyano biphenyls available at the time. There were commercials on TV for margarine, marketed as a "lower price spread" than butter, and Jim joked that ALX's LC material was the "lower-price spread" compared to the "higher price spread" made by Merck, a major producer of liquid crystal materials.

ALX's materials performed well at low temperature and sold in sufficient volume to compete with Merck. The company 3M still had a patent on the Veralight material, and that material had EBBA in it. Merck contacted 3M. A 3M lawyer called Jim and told him that if he didn't stop pushing materials into the market, 3M would prosecute ALX for infringement against their materials. Ken Marshall still remembers hearing Jim slam down the phone after that call because it was so rare for Jim to lose his temper. Shortly after, a letter arrived from 3M threatening litigation if ALX didn't stop selling in certain markets. Jim could not stomach more litigation, particularly against another Goliath, and ALX had to comply.

RECYCLING OF THEIR TOXIC WASTE

The production of liquid crystals generated alcohols and solvents as waste products. In the 1970s and 80s, many companies did illegal "midnight dumping"[46] of chemicals, but ALX wanted to responsibly dispose of their waste. Additionally, the pollution of the Love Canal off the Niagara River had caused an environmental disaster that raised awareness nationwide about the increasing pollution of the nation's waterways. The Cuyahoga River, which runs through Cleveland and Kent, was so polluted it sometimes caught on fire.

Recycling was not as common as it is now, but ALX found a company to recycle the solvents. The ALX technicians separated the short chain and long chain carbon solvents before sending them off to be purified, but the solvents came back mixed together. They couldn't be re-used when combined like this. Jim got a bright idea for disposing of the solvents. He poured them in the gas tank of the long-suffering "peanut car" that he had purchased on his European trip, to be used as fuel.

After a flood destroyed records from Ilixco that had been stored in the basement of the frame house, ALX moved to a building at 501 Gougler Avenue in Kent, just off Main Street. The building had previously housed an auto repair shop. It was built like a bunker and fireproof, which pleased the Kent Fire Department when they made their bi-annual safety inspections. The company walled off 600 square feet of the spacious ground floor and set up their first real chemistry lab for large-scale chemical preparation. Ken helped put up the walls and do the wiring for the lab, getting an electrical shock off a line during the process. The new lab had three fume hoods and benching obtained second hand. Analytical equipment went in a middle room. Jim and his secretary occupied a large front office, and Ken a smaller office at the back of the first floor.

To help out in this larger new space, Jim brought in a friend of the Fergason boys, Blake Brown. Blake lived on the same street as the Fergasons and often hung out at their house. He liked fixing appliances and tinkering with machinery. Since Jim was "better at breaking stuff than fixing it," he designated Blake as the "facility guy." Blake wired equipment in the lab and cleaned up after the chemists when they melted the plastic traps. Like Jeff and John, Blake was a high school student when he started at ALX, and arrived at work after the school day ended, sometimes with a traffic ticket in hand (he liked driving fast). He affectionately called Jim "Chief."

Jim hired a KSU chemistry student, Ken Overly, to assist Ken Marshall in the lab. Other chemistry students filled in when there was an extra order, including a woman named Nabila. Everyone who worked at ALX grew fond of her. She was a Persian Muslim who had fallen in love with a KSU student named John who was a Christian from Egypt. The two planned to marry and stay in the U.S. When Nabila's family found out, they sent her two older brothers to Kent. They knocked on ALX's door and asked Nabila's co-workers to try to convince Nabila to break up with John. The ALX workers were afraid that the brothers would kidnap Nabila, or beat up John, or both. Nabila's brothers finally gave up and went home.

SCORES OF NEW APPLICATIONS USING LCS GET UNDERWAY

To bring in more income, Jim decided to make materials on a semi-pilot scale instead of a lab scale. He retrieved the pilot scale equipment, reaction vessel and porcelain vessels he had stored in the former gas station but there was no space in their building back in town to set the equipment up. Jim looked for a bigger building to rent but then the basement at the Gougler building became available. ALX set up the pilot plant there and hired chemical engineer Jeff Schifflet to run it.

Jim always wanted good equipment and would pay a lot for it if given the chance, but Tom Murray was still bankrolling the company and kept a tight hold on the purse strings. "You had to give him a really good reason for wanting to purchase a new piece of equipment," Ken Marshall told us. When the Electronic Research Company based in Kansas City, Missouri closed, Jim bought used analytical equipment, a gastromatograph, several distometers and differential scanning equipment, all at bargain basement prices. Ken was delighted to have this equipment and used it to analyze more material.

Fred Davis, Jim's former assistant at Westinghouse, put in an order for a large quantity of cholesteric LC. Fred had founded Davis Liquid Crystals in California in 1981 to manufacture temperature-sensitive cholesteric liquid crystal items such as the mood ring and liquid crystal toys, necklaces, bookmarks, and liquid crystal stickers for kids. His company also made plastic discs with LCs in them called "Touch Me" or "Space Fidgets." One of Fred's unsuccessful product ideas was a baby pacifier with a bead of liquid crystal in the nipple. The pacifier would change color if the baby had a fever. ALX made many liquid crystal beads the size of green peas for this product, but the product never sold well because mothers were leery of giving their babies a pacifier with liquid crystals in it.

ALX was doing so well they gave bonuses to their employees. Jim landed a research contract from Becton Dickenson for a disposable thermometer that would change temperature and hold the color change after the thermometer was removed from the object. The cholesteric LC used in the thermometer had the consistency of honey. In order to properly characterize this material, ALX needed a high performance chromatograph. They eventually bought one.

Jim wanted to get in on the market for commercial alignment coatings and encouraged Ken Marshall to "put on your thinking cap and see if we can come up with a different material besides SA-72 to sell." The result was SA-74, a water-soluble polyimide-type precursor coating that converted to an insoluble polyimide after heating to around 200°C. It was first used as an LC alignment coating, but ALX also used it to coat the inside of Pyrex glass ampules that they shipped materials in to companies overseas. Liquid crystals are now shipped in aluminum containers, but at that time ALX shipped LCs in the Pyrex. One problem with this was that the LC leached ions out of the glass, raising the LC conductivity to unacceptable levels. To resolve this, Ken coated the inside of the ampules with SA-74, cured them in an oven, cooled them, put the LC in, purged out the gas and sealed the tip. This treatment made the material stable to a very high temperature and preserved its conductivity.

SA-74 was an improvement over commercial materials formulated by DuPont. The DuPont materials were already polymerized and were supplied with strong solvents, such as dimethylformamide (DMF) or n-methylpyrrolidone (NMP). The technicians dip-coated the display glass in large tanks of these solvents, which were hazardous to human health. To remove the DMF or NMP from the material, the substrates had to be baked at a high temperature. The materials then had to be disposed of, which was bad for the environment.

ALX developed a precursor material that was water-soluble and then provided a catalyst that dissolved in a small amount of DMSO (dimethylsulfoxide), which is highly water soluble. The DMSO acted as a "carrier" that helped to dissolve the catalyst into the precursor water solution. It was much safer to use large tanks of a water-based solution than large tanks of DMF or NMP. Once the techs dip-coated and air-dried the substrates, they baked them at 200°C to activate the chemistry and convert the polyimide precursor into the final polyimide. Since a high bake temperature was needed anyway to remove the DMF or NMP from the DuPont material, SA-74 was not only much better for environmental health and safety, but more cost-effective.

The other new material Ken formulated was SA-73, a soluble polyester-based material derived from PVA and a polymeric anhydride material produced by the General Aniline and Film Corp. Some of the Indium-tin oxide glass formulations used couldn't withstand the high cure temperature required for SA-74. Unless the cure was done under nitrogen or some other inert atmosphere, the high temperature changed the oxide content of the conductive coating. Like

SA-74, SA-73 also needed to be cured after coating, but at a much lower (<150°C) temperature. SA-73 was supplied as a one-bottle coating mixture with a reasonably long shelf life, whereas SA-74 was a two-part system that required mixing of the two components separately by the customer at the time of use.

ALX's materials continued to sell so well that Ken couldn't keep up with the demand in the 72-liter vessel he was using. Jim purchased a used Pfaudler 30-gallon steel reactor and they put up a catwalk to go around it. He also bought a very expensive compressor.

FUNNY, BUT DANGEROUS MISHAPS

Working with chemicals can be dangerous, and there were occasional mishaps in the lab. Ken had an assistant named Dale helping to run the equipment in the basement. One morning Dale was downstairs by himself running a reaction in DMSO (dimethyl sulfoxide) when a bad smell wafted up the stairs. Ken ran downstairs holding his nose. If the pH of DMSO falls to 4 or 5, the material has a characteristic skunk odor. Dale had stayed up late partying and fallen asleep. The pH had fallen to 5. The odor woke him up but by that time it was being exhausted to the outside of the building by the fume hood and stinking up downtown Kent.

The lab was working on a chemical synthesis that required an oxidation to be performed through the use of potassium permanganate in water. The reaction was carried out in the reaction cylinder. When potassium permanganate is mixed in a water solution, oxidization of the material causes it to heat up. Because of this, potassium permanganate must be poured into water a little at a time. One of the lab assistants dumped the chemical into the water all at once. The mixture began boiling. It shot out of the cylinder and hit the ceiling, 24 feet above. Potassium permanganate is intensely purple. The geyser rendered a large portion of the lab bright purple and made quite a mess to clean up.

The company landed a large order to produce 72 liters of highly pure cholesteryl acetate, which would bring in a substantial profit. The synthesis and purification process was accomplished by a direct reaction in toluene. Toluene is a flammable but generally easily controlled solvent. The techs started the process by boiling 40 liters of toluene in a large Pyrex flask the size of an exercise ball. A magnetic stirrer rotated in the flask to stir the solvent.

Suddenly the power went out. Heat was still being generated and the solvent kept heating up and bubbling. "What are we going to do if the power goes back on?" Ken wondered, and sure enough, it did. The stirrer turned back on and the contained vapor spouted the superheated toluene. It spilled all over the floor. Ken and his techs were afraid a fire might start and didn't know whether to run for their lives or shut the power off to the stirrer, but they were brave. They grabbed fire extinguishers and shut the power off. They had to clean up everything with an ice scraper, a tedious job to say the least. The job was a loss, but at least nobody had been injured.

Ken came up with a way of purifying p-n-pentylphenyl-p-methoxybenzoate (PPMEOB) with higher purity and less effort than it normally took. He had the system set up one day, but the distillation was taking longer than usual. By 6:30

p.m., he wanted to head home. Jim offered to stay and watch the distillation still for Ken. Ken had some reservations about this because, like most physicists, Jim tended to be "a bit dangerous in the lab," but he explained everything Jim needed to do and left. After 9 pm, Ken's phone rang. A sheepish voice said, "I think I broke the still." Jim had accidently vented air into the system, turning the inside of the still black. Ken told Jim to shut the heat off and leave it. That was the last time Jim helped in the lab.

> "Turning an invention into business is no slam dunk. A head for business is critical. If you make the best thing on earth, it doesn't matter how good it is if you only sell one."
>
> —Jim Fergason

18. ALWAYS THINKING OF A NEW IDEA

"Don't just sit there. Invent something."
—Sign that used to be on the wall of the reception room at the U.S. Patent Office in Washington, D.C.

JIM'S INVENTIVE WHEELS were always turning. As early as 1977, he began thinking about how he could increase the response time of liquid crystal displays. A disadvantage of TN displays at the time was relatively slow turn-off times, commonly 250 to 300 milliseconds. This was much too slow for the devices to be used as components in high-speed optical systems such as television displays.

The slowness was due to the way liquid crystals respond to an electric field. With the exception of a special frequency-sensitive type of LC, an electric field could not be used to turn a display off. Liquid crystals do not recognize the polarity of an applied electric field, and behave the same way whether a field is positive, negative or alternating. To turn a display off, all that could be done was switch the field to zero and wait for the internal elastic forces of splay, twist and bend to return the LC to its pre-energized state.

The speed of response of a display is inversely proportional to the square of its thickness. Making extremely thin cells to increase the speed of response would be impractical. Jim had the idea of decoupling liquid crystals in a parallel homogenous cell so that the effective layer would be the thin layer next to the surface of the cell. The speed of response of the surface mode would be an effective single layer thickness of the liquid crystal, which in turn would be a direct function of the bias.

Achieving this decoupling of layers was a design characteristic. The liquid crystal used had to be well aligned at the surface and have a very low tilt angle. It also had to have high dielectric anisotropy and as high a resistivity as could be obtained.

Before 1970, dynamic scattering (DS) displays were the only liquid crystal display type. DS relied on the bend elastic constant to turn off. Bend is the weakest of the three elastic constants, and DS displays require several seconds to turn off. The TN had a faster relaxation time because it relied on the twist constant to restore the twist structure.

Jim's idea was to create thin surface layers in a physically much thicker cell utilizing an unexploited characteristic of liquid crystals—the splay constant—to return the LC to its pre-activated state. In the central region of this new LC cell, the director was close to normally oriented and extended from the center of the cell outward to the two internal surface.

The effect of this clever arrangement was to create two active layers that were 10 to 20 times thinner than the gap between the cell's inner surfaces.

He called his new invention "compensated birefringence surface mode" or "surface mode" (SM) for short. The invention accelerated the speed of liquid crystals in digital displays," from "tenths of seconds," to "1/100,000 of a second," fast enough that the device could be used for 3D. It was a perfect component for use in certain commercially valuable applications. Today, there are three applications for SM high speed shutters currently used in commercial products. They are used in stereographic 3D glasses, welding shutters, and in the digital projectors used for 3D movies for RealD, Inc. Some computer monitors use an SM cell that is located between the viewer and the on-screen images, and covers the entire screen.

When Jim first conceived of the SM idea, he hadn't worked out all the design characteristics. In 1979, he called Tom Harsch, who was developing an LCD welding helmet for Gor-Vue, and said that he had an idea for a new invention. He used one of his favorite phrases to describe it, "It's the greatest thing since sliced bread." This wasn't hyperbole. As Tom listened to Jim's explanation of surface mode, he was struck by the genius of the idea. Out of the hundreds of scientists and engineers that had been working on TN-LC devices for the past eight years, Jim was the first to discover the surface mode phenomena. His excitement about it was infectious. He knew the SM cells would be faster than the TN cells, but there was no formula to predict the ratio. He asked Tom if he would come to ALX and make and test prototype SM cells.

Tom jumped at the offer to participate in the birth of another new technology. He resigned from his job at Gor-Vue and started at ALX. The first prototype SM cells he made were plain, continuous electrodes. He tested them to see how fast they worked, what voltages they required, and what other characteristics they had. This gave Jim evidence that the SM cells were 100 times faster than TN cells at the time. This was critical for determining that SM would be useful in taking advantage of the latency of images in human vision, i.e., why flickering film is seen as a continuous image.

The SM cell is similar to the TN in that it rotates the plane of polarization. However, a TN cell rotates the plane of polarization by a fixed 90° while the SM cell rotates it by variable degrees. A pair of linear polarizers could be used to convert rotation into the modulation of absorption. A TN cell rotates all of the visible wavelengths (or, the colors) equally by one-quarter turn, but the SM cell fans colors out by small degrees around the desired rotation value. This effect is called "dispersion" in physics. This was a drawback. Another drawback was that SM has a much narrower viewing cone because off-axis, rotation of the plane of polarization was different than the ideal one that the polarizers were configured to modulate.

Jim chose a fixed wave retardation plate as a corrective device to cancel dispersion and correct off-axis deviation. Wave plates were an integral part of Jim's concept for a practical SM system. He wanted to try to create such a system because, as he had told Tom during the TN trial, "I'm motivated to create enabling inventions because I like seeing people use the ideas I dream up."

SM DOES NOT REPLACE TN-LCDS BUT FINDS OTHER USES

Jim had not originally intended to produce displays but, after inventing surface mode, he decided to make prototype SM displays as a way to attract investors and reduce his new invention to practice. He needed more space to do this, and ALX made its third move to a building Jim rented at 777 Stow Avenue that had previously housed the Cortez Motorhome factory (NASA used a Cortez to transport astronauts from the preflight building to the launch pad). To help Tom, Jim brought in several former Ilixco employees, including Martha Dattilo, who had worked on developing the "FieldStick." They set up the prototyping operation on the second floor of the new building. The crew, which included Jim's son John, set to work.

Jim told Tom that he hoped that SM would replace the TN as a display. Replacing an invention he was famous for with a new invention would not only boost his reputation, but allow him to trump Roche and KSU, in effect winning the lawsuit. However, after several months of marketing attempts, Jim realized that the TN had become so well established in the marketplace, and so many businesses, careers, and research had been invested in it, that the SM was not going to be able to compete. After this "display" phase, Jim and Tom looked for other applications that would utilize SM's shutter capabilities. They conceived using it as an optical communicator, a 3D viewer, and a black and white to color converter for cathode ray tubes (CRTs).

They made a prototype of each of these devices. To make the optical communicator, Tom obtained a spotlight that was five inches in diameter, then fabricated circular SM cells and combined these with one linear polarizer and one retardation plate. They drove the device with an audio signal, shifting the beam of light and plane of polarization with a microphone and sound. They made a communicator and receiver, and the surface mode cell modulated the light into right hand or left hand modulated light. The detector was designed and built with a special polarizing cube consisting of two prisms connected together to make a beam splitter. Two photo detectors behind the cube picked up the two beams and fed them in an amplifier that amplified differences.

The 3D viewer used a high speed surface mode LCD to rotate the polarization of light passing through it. By putting the LC shutter in front of a CRT screen and rapidly switching the LCD on and off (or between two distinct rotations), two different images were presented to the left and right eyes. The SM cell was fast enough to switch between states without causing too much blending of the left and right images. There was also a configuration of the surface mode LCD that, in combination with colored polarizers (polarizers that only affected certain wavelengths of light, i.e. red or green), could select which colors to display. This shutter configuration could be used to create a multicolor display out of a black and white CRT.

This was especially useful because in the 70's and early 80's, it was extremely difficult to make small-sized high resolution color CRT's because of the technical hurdles of creating extremely small color masks in the CRT. Monochrome CRT's were much easier to make in a small form because there is no color mask required. Tektronix used this technology to make color displays for some of their scientific instruments. The black and white CRTs were used in cockpit displays, and have an extremely high resolution of 1500 lines per inch. Any physical means to introduce three color dots in the screen phosphor to create a color CRT requires space, which would decrease the resolution. The black-and-white to color SM color converter presented each of three high-resolution black and white images in rapid sequence that the eye put together as a single color image.

ALX's first attempts to market the SM devices show how difficult it can be to sell others on a new idea, even a great one. Jim's Midas touch for making contacts led him to the Pentagon. There Tom demonstrated the optical communicator to an admiral while Jim explained that it could be used to flood the deck of a Navy ship with modulated light that the Russians (this was still the Cold War era) couldn't see. They also demonstrated the device in front of a panel of six men from the "Black World," the CIA, who sat behind a table during the demonstration and never introduced themselves.

Next, ALX showed the color converter to a company in Milwaukee that produced optical and display devices for military applications. The company produced a very high resolution black and white TV. By combining ALX's switching red, blue and green shutters, a high resolution color display could be achieved. The company never bought the SM device, however, because it did not fit the military requirement for lifetime durability.

ALX also showed the color converter to senior managers at Polaroid. This was a livelier occasion because Polaroid had made Ilixco's polarizers and both Jim and Tom knew many people there. Fifteen Polaroid engineers donned 3D glasses and watched the demo of a 3D Rubik's cube. It wowed them and they brought in Polaroid's management, who in turn fetched 77-year-old Edwin Land, Polaroid's famous founder. But no contract resulted because Polaroid didn't want to invest in a device for TVs.

Jim's ambitions for ALX continued to grow. An article on the invention of surface mode, "Speeding up Technology with Liquid Crystals," originally published in *Inc* magazine in 1981, describes Jim's plans for ALX to tap into what Jim estimated would be a $75-million market for flat panel computer displays. "Portable computer terminals are salable—and the lighter the terminal, the better," Jim said in the article. ALX had shipped $200,000 worth of surface mode products so far, mostly engineering prototypes, to companies such as Emerson Electric, Allen-Bradley and Westinghouse Electric.

Jim was awarded his first patent for the fast-switching surface mode device in 1983. He co-owned the patent with an investor named Todd Morgenthaler. ALX built a 3D demonstration system based on a SMD on-screen modulator. StereoGraphics Corporation named this modulator the ZScreen in 1984.

The Tektronix Corporation in Oregon sent two engineers to ALX to discuss purchasing materials. Tektronix later asked for a license to the surface mode device, but Jim determined that the Oregon company had developed a surface mode cell they called the Pi cell and were using it in one of their product lines, clearly a patent infringement. His lawyer Thomas Shunk filed a case against Tektronix in 1988, but four years passed before it went to trial (which will be described shortly).

INVENTING THE ENCAP
AND ANOTHER PATENT RACE

In mid-1981, Jim announced he had an idea for encapsulating nematic liquid crystal droplets in polyvinyl alcohol (PVA). His idea was to encase the droplets in PVA (plastic). The molecules would orient to the plastic capsules, and the director would be chaotic and scatter light. In this state, the capsule would be cloudy. Application of an electric field would cause the LCs to line up with the field, allowing light to pass through the capsule. The capsule would become transparent. No polarizers would be required, and there would be no sealing or spacing issues with encapsulation.

Several days after explaining this idea, Jim, Ken and Tom used an electric mixer to whip up nematic LC and PVA dissolved in water. Whipping caused little bubbles to form and break up in beads (droplets). The solution of PVA and water encapsulated the LC beads. They poured out the beads, and let the water evaporate. It left an encapsulated liquid crystal.

Jim found a source of indium-tin oxide (ITO) coated Mylar plastic film. The team laid out a sheet of this film, spread the emulsion over it, and allowed the emulsion to dry. They placed a second tin oxide plastic sheet on top of the first one to form a sandwich. The film was opaque because the LC molecules oriented to the plastic capsules, and the director was chaotic and scattered light. They applied a voltage to the film, and in another *aha!* moment, the sandwich worked the way Jim expected. The LC director lined up horizontally with the field, allowing the light to pass through, and the film went from opaque to transparent. The bubbles of liquid crystal between plastic plates looked like "Swiss cheese" under a microscope.

The idea seemed obvious, but none of the hundreds of scientists worldwide working on liquid crystals had thought of doing this with nematic LCs before. As Jim put it, "Hindsight is always 100 percent better than foresight... That's one of the things you [inventors] are always fighting. 'Oh, that's so simple.' Well, it wasn't so simple when we started."

He called this new invention ENCAP, for "encapsulated liquid crystals." When Jim revealed this scientific breakthrough by presenting a paper on it at the 1985 Society for Information Display conference, it was recognized as a new direction in display devices. The invention is now also referred to as Polymer Dispersed Liquid Crystal (PDLC)-LCD mode, although there are some differences in structure and preparation between PDLC and ENCAP.

Tom Harsch etched a pattern on the ENCAP device to make it look like a speedometer graphic. He put dye into the device to color the liquid crystal. He and Jim took it to the General Motors research labs in Flint, Michigan, and showed it to the engineers as something they could wrap around the dash. GM was excited about the device, and asked Jim if ALX could make 100,000 of them. That's when Jim and Tom realized that the research labs at GM didn't actually do much research. They evaluated inventions others made, and let their design people decide how the inventions would be used in cars.

Jim began thinking about who could finance ENCAP development. He had no good prospects. Tom suggested their friend and former investor Jim Bell. He told Jim Fergason that it would be better to seek financing with someone he knew than someone he didn't. Jim heeded Tom's advice, and approached Jim Bell. Once again, they went in business and formed the Manchester R&D Partnership.

The date of conception was primary in U.S. patent law at the time, but Jim had been burned by the KSU experience and was anxious to file his patents as soon as possible. He and Bell met with one of the best patent attorneys in Cleveland, Armand Boiselle, who had a PhD in chemistry. As Armand listened to the description of the invention he realized it had more to do with physics, optics and electronics than chemistry and introduced the two men to his legal partner Warren Sklar. It was Friday, July 31, 1981, a day Sklar never forgot because Jim proved "the most exciting inventor I ever worked with." He was also the "best client that a patent attorney could ever have. He was a great teacher. He helped me save face when I didn't fully understand a concept by saying, 'There's no such thing as a stupid question, only a stupid answer.'"

Jim had a strong understanding of his invention and collaborated with Warren by suggesting changes to the descriptions and claims which enabled Warren to draft a strong, broad patent for him. The patent claims provide a written description of the invention, and are often compared to the property lines drawn on a deed for land. The claims define the scope of patent protection, and are the most important part of the patent application.

Sklar finished the draft of the patent application on September 16, 1981, and Jim came with Jim Bell to the office and read each page of the application as it came off the Xerox dual tape word processor. Jim had a gut feeling that he had better file the patent application that day. There was no Fed Ex then, so Warren put one of his law clerks on a plane to Washington DC. From the airport, the clerk took a cab to the U.S. Patent Office. Jim's gut feeling was correct because on October 1, 1981, a publication came out that would have restricted the claims on the ENCAP patent. The patent was awarded, and withstood 13 oppositions in Japan and two in Europe.

Business was so good that in November 1982, Ken Marshall had to put in 20 hour-days for about a week to keep up with the demand. He would go home and sleep for a few hours, then drive back to work. He ended up in the hospital with chest pains and a mitral valve prolapse, and Ken Overly had to take his place in the lab. When Ken Marshall recovered, he worked in the lab in research. He was never able to return to making materials in bulk.

19. MOVING TO SILICON VALLEY

> *"When I was getting started, I would go on airplanes and count the number of products with liquid crystal displays in them in the Sharper Image catalogs. Twenty, thirty, forty products would have liquid crystal displays. Now you go to the store and all the games, all the telephones — there are hundreds of millions of products made with them."*
> —Jim Fergason

JIM NOW WANTED to expand ALX. Lawyer Tom Murray didn't want to expand that fast so Jim made a deal with Jim Bell to buy out Murray's portion of ALX. Expenses began piling up. Jim was always looking for the next big dollar to finance what he wanted to do, but he wasn't good at managing the company's finances. He would check the mailbox every afternoon and if a check had come in, he would immediately deposit it in the company checking account.

ALX began having trouble paying the bills—and their employees. Ken Marshall didn't cash his paycheck for a week so his co-workers could cash theirs. Some employees quit, but others stayed on because they were committed to ALX's survival.

Then the IRS visited, examined the books and said ALX owed back taxes. Jim needed an influx of cash to keep the company afloat. Jim Bell promised to provide it, but when the time came he didn't have the money. He had invested in producing a play in Washington, DC even though he wasn't all that impressed by the lead actress, a young Kathleen Turner. The play bombed, closing after just one night. Turner became famous a few years later, but Bell lost his investment and ALX's finances collapsed.

Jim and Manchester R & D had no choice but to sell a license for the ENCAP technology, just as the financial pressure on Ilixco had forced Jim to sell the TN license to Roche. History seemed to be repeating itself, but Jim and Bell were now much more careful about who they approached. They showed the invention to Paul Cook, an MIT graduate in chemical engineering who founded the hugely successful Raychem Corporation in Silicon Valley in 1957.

Cook was impressed with ENCAP, and began negotiating for the license. Charles (Chuck) McLaughlin, a chemical engineer and longtime Raychem employee, handled the negotiations on behalf of his employer. Jim Bell was very involved in negotiating the licensing deal for Jim. The negotiations began before the Fergasons left on vacation to Cape Hatteras. Dora loved going to Hatteras because there was limited phone access and it was difficult for Jim to make work-related calls when he was supposed to be relaxing. Still, Jim grew restless and convinced Dora to cut their planned month-long vacation by a week so he could return to Ohio and resume talks with McLaughlin.

Bell's business acumen helped Jim and Manchester R & D get a lucrative offer for the license. Raychem was also interested in Jim's expertise because they planned to form a company to commercialize ENCAP and wanted Jim to serve as a senior scientist and technical leader. They asked Jim if he would be willing to move to northern California.

Dora and Jim talked it over. The offer was appealing, and the longer they talked, the more appeal it had. There was no love lost between the Fergasons and Kent State University. Whenever Jim wrote a paper or applied for a patent, LCI tried to elbow in. They were always trying to find out what he was up to through the KSU graduate students who worked at ALX. Beth Cunningham was a KSU graduate student in physics who spent six months at ALX doing calculations on droplet size for the ENCAP invention. She remembers William Doane, who had succeeded Glenn Brown in 1983 when Brown retired as director of LCI, asking her nosy questions about the work being done at ALX. She told him she had signed a non-disclosure agreement.

In California, Jim wouldn't feel like LCI was always looking over his shoulder. An additional draw was Silicon Valley, known as a fertile place for inventors, inventions, and technology start-ups. In addition, for the first time since his Westinghouse days, Jim would not have any administrative responsibilities and would be free to concentrate on science and technology without having to worry about marketing his inventions and balancing the books. Finally, the symbolism of moving west didn't escape Jim, who greatly admired the pioneers.

An important consideration for Dora was maintaining their standard of living. The Fergason's Kent house was worth $90,000. The same size house in the Bay Area would be much more expensive. While their three older children were grown and out of the house, Susie was just seven years old and enjoyed playing in the roomy backyard of their Kent house. Dora wanted her to have a comparable yard to play in. Jim discussed the issue with Raychem, and they agreed to rent a large house for the Fergasons. When the Fergasons were ready to buy a house, Raychem agreed to make up the difference in price.

Jim announced the closing of ALX at a company picnic in the summer of 1983. The announcement was a sad one for the remaining employees. They knew the company was going under, but had done their utmost to make ALX a success.

Everyone said their farewells. Ken Marshall got a job with the Optical Materials Group at the University of Rochester. This group had purchased materials formulated by ALX for their high peak power OMEGA laser. Tom had left six months earlier because ALX could no longer afford his salary. Ken Overly started a PhD program in chemistry.

Blake Brown packed the contents of the ALX building into a rental truck for the Fergasons and drove it out to California. Blake then returned to Ohio for the Fergason family car, an Oldsmobile, which he wired with a radar detector called a Fuzzbuster so he could speed back across the country without getting caught.

Raychem bought all of ALX's assets and equipment and the family said their goodbyes. Over their many years of working together, Jim Fergason and Tom Harsch had become great friends—and so had their wives. Jacky Harsch and Dora Fergason understood what it was like to have husbands who were obsessed with liquid crystals. Jacky remembered Dora saying in her Missouri drawl at many a party, "Now Ji-em, that's enough liquid crystal talk!"

The Fergasons left Kent in November 1983, at the beginning of a cold northern Ohio winter and headed west. They moved in to a house that Raychem rented for them in Atherton, a prosperous, pastoral town on the San Francisco Peninsula.

THE FORMATION OF A NEW COMPANY, TALIQ

The move from the Midwestern college town of Kent to the Bay area town of Atherton was a dramatic one for the Fergason family. Eight miles from Palo Alto and 25 miles from San Francisco, Atherton was an affluent town with great scenic beauty. The prices of homes averaged in the millions of dollars, and each home sat on at least an acre of land. There were no sidewalks in the town, and no commercial district. The home Raychem rented for the Fergasons, with its own tennis court, must have seemed plush to Jim compared to his family's homestead in Missouri. In California, as one of his associates put it, Jim became a sort of "gentleman farmer."

Under these comfortable circumstances, Jim began his new career as an idea person and senior scientist for Raychem. Paul Cook, Raychem's founder, thought products made with

ENCAP technology could become hugely successful. Much bigger displays could be made with ENCAP than with LCDs because all other LCD displays are liquids and the display is a flat bottle, whereas the ENCAP display is a sheet, and there is no bottle to break and nothing to leak out. Cook saw such commercial potential for products using ENCAP that he decided to create a separate company to commercialize the technology rather than incorporate the products that were developed into one of Raychem's existing product lines.

Charles (Chuck) McLaughlin, the chemical engineer and longtime Raychem employee who had negotiated the deal with Jim for ENCAP, headed this subsidiary. It was named "Taliq," a combination of the end of the word "crystal" and beginning of the word "liquid." The problem with this name was that people outside the company didn't know how to pronounce it. Chuck struggled to come up with an acronym for "ENCAP" that would be easy to say and would in itself describe the invention. He ended up lopping the first "E" off of the abbreviation for "encapsulated" that ALX had used, resulting in the acronym NCAP, for Nematic Curvilinear Aligned Phase.

Taliq set up shop in Sunnyvale to develop near-term products for the NCAP technology. Jim divided his time between Sunnyvale and the Raychem headquarters in Menlo Park, where he talked to people about his ideas for developing longer-term uses for the technology. It took him two years to establish a research and development team. Their main project was using NCAP technology to project images.

Within a year, the technical team developed several products. The first, a large liquid crystal billboard they called the "Liquid Crystal Advertiser," was a sign-changing message board that businesses could use to advertise their products and services. The text on the advertisement could be changed using a component of the invention, a device similar to today's thumb drives, that the company called a sign-loading cartridge. The cartridge was needed because wireless data communication in the mid 80's was almost nonexistent, and there were no Wi-Fi or Bluetooth standards. The cartridge, a solid state memory chip with a custom interface that Taliq created, stored up to 200 different two-line messages and could be programmed remotely. The customer could also add pre-animated graphics to their sign.

A marketing video Taliq made in the mid-1980s shows a customer typing in the message he wanted the sign to display on a Commodore Vic 20 computer, downloading the message to the cartridge, and plugging it into the sign. Taliq chose a Commodore Vic 20 because it was one-tenth the price of other computers available at the time, and used the same type of processor chip as the sign computer.

Jim's son John served as the lead technical person on both the software and hardware for the liquid crystal advertiser. Blake Brown, who Jim had invited to migrate out to California and work for him. Blake was a major contributor to the engineering and construction of the products. The product required hundreds of pages of assembly language software for both the sign controller and the editing computer interface. Following his father's tradition of teaching himself what he needed to know, John learned computer design, digital and analog circuit design, and software programming, mostly on his own.

Chuck was able to sell a large liquid crystal advertiser to a company in Taiwan that placed it on top of a skyscraper.

In California, Taliq had a pilot program with the Wendy's restaurant chain. However, within the next few years, Taliq realized the product was never going to be a rousing financial success despite all the effort they were making to market it. Years earlier, Jim and Tom had learned that watchmakers regard watches as jewelry and a watch had to be pretty. The Liquid Crystal Advertiser failed for a similar reason. While the display was readable under all lighting conditions, it wasn't as bright as eye-catching LED or light bulb displays. LEDs were power-consuming, but had the visual punch to lure customers in. The bright light solutions prevailed in the marketplace for billboards.

THE VARI-LITE VISION PANEL

Jim also developed a product called the Vari-lite Vision Panel, made with the NCAP technology. The panel went clear when an electric field was applied to it and opaque when the field was turned off. Jim was a finalist for the 1992 Discover Award for this product. In a commercial Taliq made advertising the Vari-lite as a "privacy window," two men are engaged in conversation behind a Vari-lite Vision Panel in its "on," or window phase. The men begin discussing a secret, and one of them stops the conversation to switch the panel off. It goes opaque and the men can't be seen. A longer commercial that Taliq made shows a department storefront composed of Vari-lite panels instead of glass. The panels flash different colors, and fade in and out, in synch to music, drawing attention to the clothing display in the storefront. Taliq showed this commercial at trade shows.

Taliq marketed the Vari-lite panels as an alternative to blinds and curtains, for use in showcases and greenhouses, and as a product that did double duty as both a window and wall. However, the cost of the panels, $100 per square foot including the installation cost, was prohibitive for ordinary consumers. The main customers for Vari-lite were hotels, conferences centers and corporations. John worked on the drive electronics for the demos and special installations, the biggest of which was in the pre-show area of the Star Tours ride in Disneyland, which had four large message signs and large arrays of computer-controlled windows that were synchronized to the main attraction computers. Taliq also made several large highway signs, and attempted to interest Ford and BMW (Bavarian Auto Company) in using the technology as dashboard panels in cars. It was eventually used in rearview mirror and skylight applications. However, they had more success selling proprietary NCAP displays and light switches to the exercise industry to use as panels on exercise equipment.

Vari-lite panels are still in use today. Chuck McLaughlin made a contract deal with Nippon Sheet Glass in Japan in 1985. The company is still making and selling what they call "switchable light control glass" made with NCAP technology. The glass is used in commercial buildings, hospitals, aircrafts, disco clubs, telecommunication companies, and trains in Japan and other countries.

Taliq built a factory and made a piece of NCAP film 100 meters long for Saint-Gobain, a huge international supplier of construction materials. Chuck struck a deal with Saint-Gobain and they still sell NCAP glass under the trade name Priva-lite.

In an improvement over the early Taliq commercial showed two men whispering secrets behind an NCAP window, one of the Saint-Gobain commercials shows a beautiful woman stepping behind a transparent bathroom door and beginning to slip her bathrobe off her shoulders. As the bathrobe falls to the floor, the woman flips a switch and the door goes opaque.

FORMATION OF OSI

Jim always had new ideas for inventions, so he split off from Taliq in 1987 and formed a new company called Optical Shields Incorporated (OSI), which served as an incubator for his inventions. He set up a lab in Menlo Park, and worked on surface mode technology light valves. The valves were used in 3D imaging devices, in automatic gain control on video cameras and in eye protection goggles. John Fergason began working at Optical Shields in 1988.

The company obtained funding from the Defense Advanced Research Project Agencies (DARPA) to work on surface mode liquid crystal-based devices to protect the eyes of military personnel from bright light threats, specifically laser weapons and nuclear flash blindness. Optical Shields manufactured a prototype surface-mode liquid crystal shutter in the early 1990s, and sent it to the Optical Radiation Division of Armstrong Laboratory at Brooks Air Force Base. There, the shutter was tested to see if it would offer better protection than the goggles manufactured by Honeywell. The results showed it was technically feasible to use the LC shutter in goggles. Optical Shields began selling LCD goggles to the military. When exposed to intense light, the shuttered lens instantly went opaque. The switch from clear to opaque occurred in 1/20,000th of a second.

Jim received the first of four patents for optical protection in 1988. This technology is now used in welders' masks. By 1989, Jim had 60 patents and 25 pending. He found inventing "the funnest thing I do" and didn't slow down. "What makes it all worthwhile is to try to do something that is going to create ripples throughout an industry. If you can do that, it makes for a good time," he said in a video made about him at Cal State Fullerton. He didn't invent things in isolation from each other, but rather tried to lay down one basic invention in a technical area, and then build a package of inventions around it by trying to find out which applications were going to sell and what properties would improve the patent.

Jim's sons, John and Jeff, manufactured the welding lens filter in a spin-off company they formed in 1990 called OSD Envizion, Inc. The product sold well and the company remained financially strong until the brothers sold it in 1996.

They then formed a new Ilixco and entered the 3D display business. John Fergason also worked on products using optical doubling, a technology Jim invented to increase the optical resolution on a display by fooling the eye into thinking a display had more dots. The Fergasons funded some products for a Seattle company they later bought.

JIM LOBBIES FOR INVENTORS TO HAVE LONGER PATENTS

If a patent is granted, inventors have rights to that patent for 20 years after the date the patent claim is first filed. The year

1990 was "a banner year because it was the first year that over one million square meters of LCs were made," Jim remarked. In May of the same year, his rights to the twisted nematic patent expired. On the day of expiration, Jim wore a black armband to protest the 20-year patent term. "I understand very well the reasons (for limited rights) being the case," he was quoted in an article. "It's a compromise. You don't want to hold back progress. Those useful arts become part of everyone's toolbox... but it takes a long time for some patents to reach maturity."

Jim felt the patent term should be longer because for the first half of it inventors don't have a product they can sell. Any commercial success an inventor has trickles in years after the inventor comes up with the idea. Inventors often have a great idea and concept, but can't execute it for various reasons, including the complexity and cost of developing an idea into a sellable product. With the LCD, Jim never made much more than $1 million off an invention that by 2003 was generating billions of dollars in the global economy.

The twisted nematic continued to be Jim's most famous invention, and he was proud that it "produced more permanent jobs than any job program in recent history." However, it pained him that the manufacturing of these displays almost all took place overseas, mainly in Japan. There were few jobs in LCD manufacturing in the U.S. Jim told a *Forbes* reporter in 1990 that, "U.S. industry just didn't have the vision to see the market for liquid crystal displays. We flubbed it badly. For the country, it's a damned shame." Jim elaborated on why the market got away from the U.S. "U.S. companies didn't anticipate what the liquid crystal would spawn," he told a reporter.

"They thought it would be a 'flash in the pan,'" adding, "I have a lot of respect for the Japanese. I have every reason to respect their ability. On the other hand, I'd really like to beat them at their own game if I could."

Jim was a member of the International Society of Optics and Photonics. The CEO, Eugene Arthurs, testified in a Congressional hearing in 2009 on China's industrial policy and its impact on U.S. companies, workers and the American economy. He recounted a conversation he had with Jim on why there was no LCD manufacturing in the U.S. They discussed why almost all the jobs related to liquid crystal display (LCD) technology were in Asia, although most of the invention was in the U.S. Jim pointed out that the facile response, lower labor costs, was not correct, as the labor content of the typical LCD was tiny. His view was that the lack of jobs in the U.S. was the result of major U.S. corporations' unwillingness to invest "substantially and patiently" in liquid crystal manufacturing. He also commented on the lack of visionary leadership in U.S. blue chip companies, and in the particular case of LCDs, attempts to protect older display technology (at that time owned by then substantial U.S. suppliers to the consumer electronics markets).

Jim came into his own as a fierce advocate for the rights of independent inventors after moving to California. He still had one more battle to fight for his own rights when the case he had filed four years earlier against Tektronix finally went to trial in October 1992.

20. THE TEKTRONIX LAWSUIT: A SET OF PIRATES

> *"I have been so constantly under the necessity of watching the movements of the most unprincipled set of pirates I have ever known, that all my time has been occupied in defense, in putting evidence into something like legal shape that I am the inventor of the Electro-Magnetic Telegraph."*
> —Samuel F.B. Morse, letter to his brother, April 19, 1848

RECALL THAT JIM'S lawyer, Thomas Shunk, had filed the case against Tektronix in 1988, for infringement of the surface mode cell patent during the ALX era. There were several reasons why it took four years for the case to go to trial. One was that Tektronix wanted to overturn the patent and requested that the U.S. Patent Office conduct a reexamination. This is a formal procedure that required Tektronix to provide an explanation of why the claims in the patent should be invalidated. Tektronix had to either provide evidence that identical or near-identical technology existed before the patent was filed, or that the technology was publicized in some way before the filing date.

The case was halted during the two years it took for the patent office to conduct the reexamination. To counter the request, Jim as the patent owner and his attorneys had to provide evidence supporting the claims in Jim's issued patent. In the end, the patent office determined that the patent had been correctly issued to Jim.

The trial was also delayed because Tektronix filed nine motions for summary judgment, a claim essentially contending that there are no facts in the case that can be tried. Every court has slightly different rules, and this particular court had no limits on how many motions could be filed. The day the judge provisionally granted the motions was a dark one for Jim and Thomas Shunk because if the judge finalized his decision, it would have prevented the trial from proceeding. To Shunk's great surprise, however, the judge reversed his decision in a hearing.

The trial was held in a court in Akron, Ohio in front of a jury composed of the usual cross section of society. One of the jurors worked nights as a jeweler and kept nodding off during the trial, but the judge decided not to kick him off the jury. The trial lasted six weeks, which is long for a patent trial. Jim stayed in an Akron hotel on the Cuyahoga River for the duration, with the exception of a stay of several weeks at Tom and Jacky Harsch's home near Kent, where he got a reprieve from the stress and ate home-cooked meals among friends. Winter had hit Ohio early and hard. During the entire trial, the weather stayed icy. Jim got terribly homesick. Dora could only visit for a short time because she was taking care of Susie. Jim was able to travel to California for just a few days over Thanksgiving.

Thomas Shunk was assisted by Jay Campbell, a young lawyer with an engineering degree who did the grunt work and wrote many of the motions that were filed. Warren Sklar knew the most about the technology and provided counsel.

Jim and his lawyers maintained that Phillip Bos, a Tektronix scientist, had stolen Jim's idea for the surface mode cell after Tektronix employees heard Jim talk about the cell at a conference. Tektronix claimed that Jim had not invented the cell, that he had misled the U.S. Patent Office and that his cell was different from the one Bos had invented.

Wrangling with Tektronix made the case with Hoffmann-La Roche look like a walk in the park. The Tektronix lawyers were "vicious, and over the top." They maligned Jim's character and portrayed him as dishonest by saying that he claimed to have a PhD. They took depositions from everyone who had worked with Jim, including his son Jeff.

Tektronix offered Alfred Saupe a fee to testify, but when he didn't give them the testimony they wanted they didn't pay him. Ken Marshall was questioned for eight hours, with only a break for lunch, and Tom Harsch for three days. Jim's deposition for Tektronix lasted a marathon six days.

Tektronix went after Shunk in front of the jury, rhyming his last name with "skunk" and calling him a sneak. It was so bad that Shunk had his wife sit in the courtroom one day to try to read how the jury was reacting to Tektronix' attempted character assassination. Tektronix' tactics backfired. Dora ate lunch at a hotel during her short visit to see Jim and overheard some of the jurors talking at the next table remarking that they disliked the name calling.

Shunk had to take very complicated science information and explain it in a way that was understandable to jurors with just a college degree or a high school education. One of the best ways to do this was through demonstrations. Jim had made 3D movies that he used to demonstrate the surface mode invention. He set up a viewing of these movies for Warren Sklar and his wife. They put on head-mounted displays which gave them a left eye and right eye image. Combined, the images appeared in 3D. The first movie he showed them was of children on a merry go round, grasping for a brass ring. The children reached towards the viewer for the ring. The second movie was of children flying kites, and the kites looked like they were flying towards the viewer. The third movie gave the viewer the experience of sitting in a flying plane. This experience was so realistic to Warren's wife that she recalled leaning left, and then right, to balance the movement of the plane.

Shunk decided to show the jury the 3D movie of the kids flying kites. The jury donned special 3D glasses. After the trial that day, a reporter asked if he could see the movie. He put on the glasses, and Jim did too. That's when Jim realized that the images were switched. Instead of the kite flying toward the viewer, it flew away. The lawyers realized the switch was thrown the wrong way. They asked the judge if they could re-show the movie to the jury, and the judge said yes.

One of Tektronix' claims for invalidating the patent was that Jim had already demonstrated the surface mode invention at a conference in Cherry Hill, New Jersey. This constituted prior disclosure. Jim had not filed for the surface mode patent within the year after he disclosed the invention, which was a requirement for patenting.

However, Jim had an ace up his sleeve. He testified that when he attempted to demonstrate the invention at the New Jersey conference, the demonstration completely failed because the power cord wasn't long enough to reach the closest outlet. He didn't worry about filing for the invention within a year of disclosure because *there had been no disclosure.*

Tektronix tried to do the exact same demonstration in the courtroom that Jim had done at the conference. Ironically, it also failed because of the same problem—the extension cord wasn't long enough to plug into the outlet. Instead of proving Tektronix' point that the demo had occurred, Tektronix literally ended up demonstrating that the demo had not taken place, just as Jim said.

Philip Bos from Tektronix did not testify in person about how he invented the Pi cell, an absence Shunk told us he found "surprising." Instead, excerpts from Bos' deposition were read to the jury. Bos testified that he had attended a lecture Jim gave which sparked his thinking and led him to what he believed was a different solution to the problem of getting liquid crystal light shutters to rapidly switch.

Tom Harsch also testified in front of the jury. When Jim made the deal with Raychem, Tom had asked if he could invest in it and was giving a 5 percent stake. The Tektronix lawyers knew that Tom had part ownership in the Raychem deal and assumed he had been bought, was colluding with Jim, and had made up the story about how they reduced the surface mode device to practice. They kept asking Tom questions to prove he and Jim had lied. But he hadn't, and it was evident the jury believed him.

At the height of this stressful trial, when Jim's reputation was on the line, Jim asked Shunk's assistant Jay Campbell to give him a ride to a store. Jay had mentioned to Jim that he had a young son and right before the trial, Jim had watched *Beauty and the Beast* with one of his grandchildren. Jim bought a video of the movie at the store and gave it to Jay to watch with his son. Jay never forgot Jim's thoughtfulness and still has the video.

THE JURY DECIDES IN FAVOR OF JIM

After the trial ended, the jury began deliberations. Jim and Warren Sklar couldn't stay for the verdict because they had to fly to Munich. Jay called Warren in Munich with the news that, after four days of deliberations, the jury had found that Tektronix had willfully infringed on Jim's patent. "That was one of the best calls I ever made," Jay said. It was late at night, but Warren called Jim's hotel room to give him the good news. The telephone woke Jim up, but Jim was too excited to go back to sleep. The men decided to meet in the hotel bar for a celebratory drink. The bar was so packed they couldn't get near the bartender to order. A young man noticed their predicament and asked them what they wanted. "We want a beer to celebrate a patent win," Warren told him. The young man fought his way through the crowd and came back with the beers. Warren tried to give him money, but he wouldn't take it.

Jim and his lawyers were scheduled to meet with the judge after the Munich trip to hear what damages he would award. Because the jury had ruled the infringement was willful, Jim could be awarded damages of up to $15 million at the discretion of the judge. Tektronix offered to settle with Jim and his investor Todd Morgenthaler. During the trial, Jim had requested damages of $10 million, about 10 percent of the $95 million Tektronix said they had made on the sales of the Pi cell, and Tektronix now offered this. If Jim agreed to it, Tektronix said they would not file an appeal. Between the trip and the trial, Jim was so tired that he didn't want to hear what the judge had to say about the damages, and agreed to the settlement. Tektronix also paid Jim a license fee to continue using the invention in their products.

In the trial against Hoffmann-La Roche, Jim had certainly not "won" very much money. Battling Roche had wiped him out financially, been a great strain on his family, and damaged his health. However, he had won something that was more important to him than money—the recognition that he was the inventor of the TN-LCD. The award in the Tektronix trial was monetarily much greater, and it would seem to casual observers that Jim's pursuit of the lawsuit was about just that: money. However, Jim was not the kind of person who pursued wealth to the exclusion of anything else. Of greater satisfaction was knowing that billions of people were using his TN-LCDs, surface mode, and NCAP inventions.

21. MR. LIQUID CRYSTAL

> "I can still picture Dr. Fergason standing before a packed house of aspiring inventors [at an Independent Inventor Conference] telling his story of hard work and success and the audience reacting as if he were a rock star. In fact, Jim Fergason is much more than that. He is an American hero."
> —Nicholas Godici

INVENTING WAS IN Jim's blood and even though he was 60 in 1994, he wasn't about to slow down. He had been making a good living off his inventions for ten years and with the Tektronix settlement alone, could easily have retired. But, as Jim told one reporter from the *San Jose Mercury News*, he found "nothing more fun than being an inventor. Maybe it's the same thing that makes other people paint. It's very creative. I'm doing things for the first time—things no one else has done—and it gives me a hell of a high.

Jim had certainly experienced the lows following the highs. The battle with Hoffmann-La Roche motivated him to continue advocating for independent inventors. His admiration for George Washington Carver, Edwin Howard Armstrong, George Westinghouse and Nikola Tesla also spurred him to help others in the field. Each of these inventors started from a humble beginning, became a successful independent inventor, encountered great career difficulties and eventually received recognition. Their triumphs, perseverance and grit inspired Jim, and their tragedies—Armstrong had committed suicide in 1954 in the midst of a prolonged patent battle over his inventions and Tesla died penniless—made him all the more committed to helping other independents succeed.

One way he did this was to share his knowledge and experiences. "I'll say something as a kind of entrepreneur," he told a studio audience of young inventors in 1999. "When you do these things [inventing], don't expect them to be easy. They never are. Everything's going to be hard...If the invention is obvious, getting it to market is not obvious."

Not only did bringing an invention to market take time and resources, it took an understanding of what Jim called the "hooks and crooks" to inventing. With two lawsuits, four startup companies of his own, and an invention career of 40 years behind him, Jim understood the legalities of patenting as well as the ins and outs of protecting one's intellectual property. Absolute novelty, he told the audience of young inventors, was a requirement. "Nobody can find out about your invention before you file your patent... if we come up with an invention today and talk about it we're in a little bit of trouble," as inadvertently disclosing an invention was a real danger that had destroyed many a patent. He gave the example of the windsurfing invention, and the major lawsuit over whether it was an invention or not. "Way in the background of a photo published in *Popular Science* magazine, someone was standing with a windsurfing rig," Jim explained. "That constituted prior disclosure and destroyed the patent for the invention."

Jay Campbell said of Jim that he was, "as honest and helpful as could be." For a while, Warren Sklar went to industry conferences with Jim and followed him around to learn more about Jim's inventions so that he could write better patents. Warren was struck by how often other scientists came up to Jim and asked for help solving their problem. Jim willingly shared his knowledge, and often came up with solutions. Sometimes, Warren felt that Jim came up with ideas that he might have kept to himself because they were potentially patentable.

Jim's great personal warmth made him an effective ambassador for independent inventors. He eagerly shared his process of invention. He wanted the public to understand his philosophy: "The inventive process comes down to how you examine a problem. You need to look at a whole problem. You need to look at a phenomenon and say, "what's this good for? What question can I ask? How can I do this better?"

AND MORE INVENTIONS

Throughout the 1990s and beyond, Jim kept up a dizzying pace of inventing. In the early 1990s, he invented a dynamic contrast technology called System Synchronized Brightness

Control (SSBC) to improve image quality, specifically image detail and contrast for displays, particularly those for televisions. At the time, a TV with this component couldn't be built because the technology wasn't ready. The idea seemed like a minor invention. Now the device in modern TV's dramatically increases the dynamic range, the range of tones from brightest to darkest, on a TV display and the color quality of video and photographic images on electronic displays. Another SSBC application is a scene-dimming shutter that is controlled by signals embedded in the video stream that close the shutter to darken the scene when, for example, a movie character enters a tunnel. The shutter then opens fully when the scene moves into bright sunlight. SSBC shutters are also used in night scenes.

Nonetheless, in 2000, Jim called Chuck McLaughlin and asked if Chuck thought any of these older patents could be sold. Chuck thought they could, and became the licensing agent for Fergason Patent Properties, a company Jim started in 2001. Jim gave the company a portfolio of patents to license. The SSBC patents turned out to be a valuable family of patents. Chuck licensed the invention to almost every major TV manufacturer in Europe and Asia, including Sony, Samsung and Sharp.

Jim was never alarmed when an experiment failed. He regarded every failure as an opportunity to learn—and possibly invent something else. Warren Sklar witnessed this when he visited Jim in California. His lab technician told Jim about a problem that kept occurring. She made an LC cell again and again, but when she applied voltage to it, it burned out. Jim went across to her lab and observed this for himself. He said to the tech, "Before we try to solve this problem, let's see if we can use this effect to our advantage."

Jim urged inventors to research and read widely in their field. "One of my best collaborations has been with the past," he said to an audience of young inventors. "I find that when you go back over things and look at what people have done before, a lot of people miss things, and that's been one of my best sources of inspiration. It has also been a source of inspiration to see how smart some of these old guys were. Studying their work gave me clues about where to look that had been overlooked for years...a lot of my inventions I owe to the people that came before me. When I'm reading I'm building up knowledge and research so that when I turn back to inventing, I have all this new information...You have to keep abreast of things."

Jim spoke out against laws that threatened the rights of independent inventors. He served on the board of the Intellectual Property Owner's Association in Washington DC, an organization that educates the public on intellectual property rights and advocates for laws that support those rights, such as keeping the cost of patenting down. Jim traveled to DC often in the 1990s, lobbying members of Congress to support patent protections for independent inventors. On at least two occasions, he testified before Congress on a proposed patent bill.[47]

He was accomplished in this role. He was an intellectual, who could talk knowledgeably about art, books, movies, classical music and politics, and had strong opinions about everything. Such interests weren't unusual among physicists before World War II, when the profession "was a haven for eccentric, adventurous intellectuals who mixed art and philosophy (as well as politics) with their science," but it was rarer after the war, when the profession became high-status and physicists could make a lot of money.

The executive director of the Intellectual Property Owner's Association, Herbert Wamsley, said of Jim that "he's a creative genius who doesn't fit the stereotype of the inventor 'nerd.' He's also an engaging, articulate conversationalist." The 'nerd' stereotype is "largely false," Wamsley remarked to us when we interviewed him. "Most successful and prolific inventors have business judgment and people skills. Jim had a good sense of humor, traveled widely, and was worldly," he told us.

Jim enjoyed telling jokes. His favorite described the way that physicists like him felt about chemists, mathematicians, and others who he felt were too cerebral. "A chemist, an engineer, and a mathematician arrived in a large city to attend a scientific conference," his joke began. "At their hotel each was given a room, an efficiency suite. After dinner each went up to his room, settled in, and went to bed."

Jim would then relate how late at night, the chemist was awakened by the smell of smoke in his room. A fire had broken out near the entrance door. The chemist jumped out of bed, went to the door and examined the fire. He noted the chemical nature of the materials being consumed, as well as the temperature and rate of spread. He ran back to his bed, made a few calculations in a notebook, then to the kitchenette where he measured out the right amount of baking soda, and other chemicals he thought would help, ran to the door and threw them onto the fire. This put it out and he went back to bed.

A fire broke out in the room of the engineer. He watched it burn for a moment, ran to the bathroom where he found a cleaning bucket, ran to the kitchenette, filled the bucket with water, ran to the door, threw the water on the fire, put it out, and went back to bed.

A fire broke out in the room of the mathematician. He woke up and studied it for a while. Then, he picked up a notebook which he always kept on his bed stand (to write down good ideas that might come to him in the middle of the night), made a few calculations and muttered, "A solution exists." And went back to sleep.

Jim loved to fish and stroll on the beach at Cape Hatteras, NC, where he and Dora owned a beach house. He was a kind, patient and attentive father, except when he was working on a problem. At those times it was useless to ask him a question, something Jeff found frustrating as a kid. He enjoyed talking with his children about history and science.

When Susie was in high school, Jim came home with a device that he wanted to show off. He told Susie to get her Walkman, and he plugged a small device with a reflector on it into the headphones port. He then placed it in the bathroom all the way down the hall from the family room — about 20 feet away. He had another device that looked like a laser that he attached to a speaker. He shone the laser light down the hall and the music from Susie's Walkman started playing out of the speaker. Susie had a friend over that day, and she was amazed at the device. It was right after the movie *Beauty and the Beast* came out, and in the movie, Beauty's father is an inventor. Susie's friend told everyone that Jim was like the father in *Beauty and the Beast*.

LOBBYING TO CHANGE PATENT LAW TO FAVOR INVENTORS

In 2001, several members of Congress proposed changing the United States' first-to-invent patent law to a first-to-file patent law. While proponents of the law, which included Microsoft, GM and IBM, argued that the change would both bring U.S. patent law in line with the first-to-file patent laws in other countries and encourage invention, opponents argued that the new law would harm independent inventors and smaller companies, forcing them to file quickly or risk losing their intellectual property rights. Larger companies with on-staff lawyers would have the means to get patents filed quickly.

Under the first-to-invent law, inventors had a grace period of about a year to file a patent application without risking the loss of their intellectual property. This gave them time to determine whether an idea they had for an invention would work. Under the new law, the grace period remained, and inventors could publicly disclose their idea before filing for a patent, but the definition of the grace period changed, making it riskier to disclose before filing. The USPTO's proposed rules for the "First-to-File" system, published in the July 26, 2012 issue of the Federal Register, "clarified that most disclosures by third parties will continue to be treated as prior art even when a third party disclosure is preceded by an inventor's own public disclosure."

Under the old law, your invention had to be "enabling" to obtain a patent, while the new law encouraged the patenting of concepts. A conceptual invention is different from an enabling one. Jim always said, "an idea is not the same as an invention." Under the old law, you had to demonstrate the device could work to obtain a patent.

Jim strongly opposed the change in the law, and spoke out against it often. Europe had a first-to-file system and for that reason Hoffmann-La Roche had been able to obtain the Swiss patent for the TN-LCD back in the 1970s, whereas their claim to the invention did not hold up in U.S. court, which then had a first-to-invent law. It was more difficult to obtain a patent under the old law because inventions had to be enabling ones. As Tom Harsch put it, "the previous American law supported inventions that were proven to work."

After Jim won the Lemelson-MIT Prize in 2006, he expressed his concern about the next generation of inventors because of the proposed change in law. He said, "This law... has no provision that will actually improve the patent law or will strengthen patents but will weaken them considerably... To sum up the overall problems with this new legislation: it makes it harder to get the patent, it makes it harder to license the patent and it makes it harder to make money over the overall license of the patent. And it makes it easier to steal [someone's invention]."

Independent inventors were vital to the U.S. economy, Jim felt, and by protecting their rights, the U.S. would support economic growth. "As a country our economy is depending more and more on other things than manufacturing," he said. "Manufacturing is no longer a big part of the American economy. In this instance the individual inventor who brings something new to the table is able to build our economy and provide high quality jobs. It should be made easier for him to obtain a patent rather than harder. It should be easier for him to protect his invention rather than harder... It's clear that inventions are the engine of our modern economy."

At Westinghouse, Jim initially had to scrounge for support for his liquid crystal research, and at Kent State, he had been stymied by the academic bureaucracy and Glenn Brown's emphasis on basic rather than applied research. In his own companies, he made the decisions. He could push an invention forward, as he had done with the TN-LCD or abandon it if he recognized it wasn't going to work, or wouldn't be profitable. "I have always been willing to let an invention go if it doesn't look good," he said in a 1993 film about inventors. "The result is I have a very high batting average of issued patents." Economic considerations affected his decision-making process. Because of the cost of patenting, "you have to decide if there's really a market for the invention."

JIM GIVES BACK TO HIS SCHOOLS

It was very important to Jim to help advance science education in the U.S. After he moved to California, he got a call from his old high school in Carrolton, Missouri. They were canvassing for donations to refurbish the chemistry lab. The lab had never been updated. The students still sat at the same stone chemistry benches that Jim had sat on, and used antiquated equipment. Jim was appalled, but not really surprised. The district was tiny. They didn't have funds to support science education.

Jim never hesitated to buy the best equipment for his own research, and he didn't hesitate now. He donated enough money to completely outfit the chemistry lab. To further encourage the students to use their newly outfitted lab and to consider a science career, he and Dora decided to fund a scholarship awarding $2000 to a graduating Carrolton High senior pursuing a four-year degree in science or health.

Without the partial scholarship the University of Missouri had awarded Jim, he may not have been able to go to college. In 2001, he and Dora created the James L. and Dora D. Fergason Fund for Excellence in Physics at the University of Missouri.

A LIFETIME OF ACHIEVEMENT AND MANY AWARDS

Jim's stature as a prolific and important inventor continued to grow. Dora ran out of wall space to hang up Jim's awards. The first award Jim received was the Industrial Research Magazine's IR 100 award, for one of the 100 most significant inventions of the year. Other important awards he received included, in 1986, the Francis Rice Darne Memorial Award from the Society for Information Display for outstanding technical achievements and contributions to the display field. In 1989, he was recognized as Distinguished Inventor for his nonlinear optical eye protection with sub-nanosecond response by the Intellectual Property Owners in Washington, DC. Other awards included the 1989 Laurels Award from Aviation Week and Technology, and the 1990 Quiet Hero Award from Electronic Design. In 1996, the American History museum of the Smithsonian Institute honored Jim for his role in the development of the quartz watch.

Jim Fergason and Tom Harsch at Jim's induction ceremony into the Inventors Hall of Fame, 1998

One of the honors that meant the most to him was his 1998 induction into the National Inventors Hall of Fame. Ira Flatow, NPR's *Science Friday* host, emceed the induction ceremony. This glamorous, black tie affair was held on September 19, 1998, at Inventure Place in Akron, Ohio, then the location for the Hall of Fame (which has now moved to the campus of the US Patent and Trademark Office in Alexandria, VA). Jim's entire family attended, as did Tom and Jacky Harsch, Thomas Shunk and his wife, and other friends.

In the ceremony, modeled on the Academy Awards, a biographical film clip of each inventor honored was shown to the audience before the inventor appeared on stage. In the film, Jim described himself as a "country kid who liked to blow things up." He also described how intrigued he was with the optical properties of liquid crystals. "Bats are ear directed. Dogs are directed by scent. Humans are eye directed. Anything that allows information to be transformed into an optical image is important." After the film of Jim was shown, he burst out on stage like the Wizard of Oz from behind a giant curtain. With an impish grin, lifted up his left arm and gazed at his liquid crystal watch.

In February, 1999, Jim was recognized by Missouri University as a distinguished Arts and Science Alumni. At the awards ceremony the Dean presented Jim not only with an attractive commemorative plaque, but, curiously, a framed copy of his college transcript. Jim hung both the plaque and the transcript on his office wall and chuckled every time he looked at the transcript.

HIS HEART KEPT TICKING

That same year, Jim began feeling increasingly fatigued. Although he enjoyed walking on the beach (where, ever the animal lover, he gleefully fed the seagulls), and was an avid sports fisherman, he would rather read a good science fiction novel in his spare time than exercise, and he loved to eat. On the family farm in Missouri, the Fergasons never went hungry, even though they didn't have much money. Farm work was hard, and the meals correspondingly hearty. Jim was used to eating big meals.

Duane Werth, who worked for Jim at Ilixco, said that, "I still consider Jim the most intelligent person I ever worked for." He observed that Jim "needed to eat to keep his mind going. I remember him saying one time that he needed to run out and get something to eat so he could think, and he was very serious about that."

The Fergasons enjoyed entertaining and often hosted parties in the large house they had built in Atherton. Dora always prepared delicious food in her custom-built kitchen. She was such a good cook that many a friend or colleague of Jim's made one excuse or other to drop over at dinnertime, something that both amused and annoyed her. After Jim's heart attack in 1976, he'd attended cardiac rehabilitation classes and knew how important it was to keep his weight down. Dora cooked lower fat foods. It had been 24 years since his heart attack, and Dora was concerned about his fatigue.

His cardiologist then diagnosed him with blocked arteries. So on September 8, 2000, Jim underwent a quintuple bypass. The procedure, an open-heart surgery, requires the surgeon to split open the sternum. "They cracked me open like a crab," Jim said to daughter Terri when she came to visit. For the surgery, the surgeons had to harvest a leg vein that went all the way from his ankle up his entire leg. The surgery left a huge incision going up his leg. Years earlier, when the Fergasons lived in Pittsburgh and Jim worked at Westinghouse, he awoke one night in terrible pain and had emergency gall bladder surgery. In those days, they made a horizontal surgical cut across the entire abdomen. Terri joked with her father that with the two scars, he was like James Bond when Goldfinger tried to cut him in half.

WRITING HIS OWN BOOK

Recovery from a bypass takes about three months, but Jim felt well enough in 2001 to join Dora on an around-the-world cruise. On the ship, he began writing his life story. Roche had appropriated the creation story of the TN-LCD, and Kent State University still claimed that Jim had invented the device while he was in their employ. These claims chafed at Jim, and he'd been thinking about writing his own book to refute them. In addition, people always asked him how he came up with the ideas for his inventions, and he wanted to write a book to explain that. It might have been the scare to his health that prompted him to finally begin.

In the manuscript, which he entitled *Birth of a Technology*, Jim included a text box listing some of what he called, in the neutral language of science, "factually incorrect statements" about the invention of the TN-LCD, followed by his rebuttal of these statements. For example, he presented this quote from *Crystals that Flow*: "... Fergason subsequently filed a patent in the United States in April 1971 which describes the twisted nematic display, but Hoffmann-La Roche had the edge and bought Fergason's patent rights."[48] Jim's rebuttal read: "Hoffmann-La Roche purchased the Ilixco patent portfolio because they had a financial "edge" but certainly not a legal edge. Their patent was rejected in the U.S. and Germany while the Fergason patent was issued."

Jim worked on this manuscript over the next few years. He wanted to write it himself, but realized he didn't have the writing skills to pull it together and asked for help from a colleague and friend, Arthur Berman. Berman was helping Jim work on a book about cosmology and relativity, but work on the biography soon took precedence.

JIM'S LAST YEARS

Four years after his bypass surgery, in 2004, Jim developed a dry cough that wouldn't go away and pain behind his shoulder blades. Just before Thanksgiving of 2004, x-rays showed lung cancer. Jim had never smoked. However, he was around second-hand smoke all of his life.

Jim and Dora were planning to travel to Susan's house in Seattle for Thanksgiving with all of the children and grandchildren, and Jim still wanted to go. He told Dora that there wasn't any reason to hang around home and that he wanted to see his girls. He didn't tell anyone about the diagnosis at Thanksgiving dinner. The rest of the family didn't find out until Jim began treatment after the holiday.

He wasn't a candidate for surgery on his lung cancer because of the bypass surgery he'd had four years earlier, so the cancer was treated with chemo and radiation. The treatments made Jim's skin so raw he had to wear loose clothing. At the grocery store one day he ordered cold cuts at the deli, and the clerk asked him if he was homeless and could pay. The radiation shrunk the cancer, and it looked like it was going in remission. Jim carried on with his life, and with inventing. The prizes kept rolling in.

He received the most prestigious, the Lemelson-MIT award, in 2006. It comes with a $500,000 cash prize, the largest given to inventors in the U.S. Jim continued his philanthropy by donating all of this award money to fund science scholarships. Half went to the Excellence in Physics scholarship fund he and Dora had established at the University of Missouri five years earlier. He donated the rest to LCI to establish the Alfred Saupe Scholarship Fund for graduate students. He also gave large donations to several hospitals to fund cardiac units. Two-hundred thousand dollars went to the Sequoia Hospital Foundation and $500,000 to the Sequoia

Hospital Building Fund. He and Dora were lead donors to The Children's Heart Center at the Lucile Packard Children's Hospital. Jim and Dora also donated to the Palo Alto Medical Foundation for the purchase of an imaging machine for use in cancer treatment. Jim didn't do any of these donations for the recognition. As his son John put it, "my father didn't want his name under lights. He tried to do some good with his wealth, and help out people in some way."

In Jim's later years, he enjoyed a comfortable life. Although he had never been interested in sports, he began going to professional hockey games. On a whim he bought season tickets to the San Jose Sharks, and John went with him to many of these games because Dora didn't like to go. For as long as Jim had those tickets he never sold any of them—he gave them away if he couldn't go himself.

In 2007, Jim fell ill again, and the doctors found that he'd had a recurrence of lung cancer. The IEEE had only recently informed Jim that he was going to be awarded the Jun-Ichi Nishizawa medal for "pioneering development of the twisted nematic liquid crystal technology." As ill as Jim felt, he did not want to miss the awards ceremony, but he couldn't travel by regular plane. Jim chartered a private plane to fly the family to Quebec for the ceremony. Terri remembers her surprise at seeing the program, for her father was one of three people receiving the award. The co-recipients were Wolfgang Helfrich and Martin Schadt. Her parents had shielded their children from the controversy over who had invented the TN-LCD, and she didn't even know that Helfrich and Schadt claimed the invention. Accepting the award was a great pleasure for Jim, but it also frustrated him that Helfrich and Schadt were co-recipients.

It was time to finish his book manuscript setting the record straight. Despite his poor health, he and Arthur Berman kept working on the manuscript, and by October, they had completed it. Jim knew it would need editing, but he wanted to get it published, and sent it to a literary agent in New York. They came to an agreement, but Jim was too ill to work any further on the book. The cancer spread to his back. He rallied enough to spend one last Thanksgiving with his family. He could still move around, but only with a walker. The holiday was a tremendously sad one for the family. Terri's youngest son Tim started reading his grandfather's book manuscript. Soon after Thanksgiving, Jim became bedridden. On December 7th, Terri, who was at his bedside, told him that Tim had finished reading the manuscript. Jim brightened up. "Tim read my book!" he said to her excitedly. "Did he like it?"

"Yes, he liked it a lot," Terri said and Jim gave her his big, warm smile. He died the next day, December 8, 2008.

FOREVER MR. LIQUID CRYSTAL

Jim was the "Johnny Appleseed of liquid crystals." He seeded the entire liquid crystal industry, in no small part because, as Tom Harsch put it, "Like Michael Farraday he believed in the power of scientific demonstration" and was a "master at demonstrating his inventions." LC scientist Allan Kmetz considered Jim a "creative giant in what was the evolution of what is now a huge industry." Fred Kahn wrote that Jim was a "leader in his field...extraordinarily prolific and always a gentleman."

Jim said of liquid crystals in an interview, "I found liquid crystals—or they found me."[49] He compared his many inventions in liquid crystals to an opus. "All distinct inventions related to one central theme." He was often called a pioneer, "but I identify more with the mountain men who went in first, on foot or horseback with primitive equipment, exploring the new country and marking trails for others to follow."

Ten years before Jim died, Warren went with him to the patent office in Washington D.C. to interview a patent examiner. Warren had been a patent examiner himself and still knew some people at the office. He was walking with Jim down the hall and a former co-worker approached, pointed at Jim and asked, 'Do you know who that is?"

"Why yes,' Warren said, "he's Jim Fergason," and the man said, "Around here, he's known as Mr. Liquid Crystal."

ENDNOTES

1. David Dix, "Along the Way," *Ravenna Record Courier*, 1998.

2. "The Cary Family," unpublished document provided by Emily Cantrell. The Cary family genealogy in this document is from "Biographical Sketches of Citizens of Carroll County, Missouri" and "Illustrated History of Carroll County, Missouri — 1876."

3. Steve Kaufman, "Prolific Inventor Loves Job," May 8, 1989, *San Jose Mercury News*.

4. Ben Vester, an oral history conducted in 2009 by Sheldon Hochheiser, IEEE History Center, New Brunswick, NJ, USA at the National Electronics Museum, Linthicum, MD, USA. http://www.nationalelectronicsmuseum.org/BenVester.shtml

5. Max Garbuny, "In the Clutches of the SS," essay originally published in *We Shall Not Forget: Memories of the Holocaust*" by Carole Garbuny Vogel, http://www.recognition-science.com/cgv/clutches.htm

6. O. Lehmann, "On Flowing Crystals," Zeitschrift fur Physikalische Chemie 4, 462-472, 1889.

7. Anchor Optics Catalogue, X490A, section titled "Polarized Light Application," published in 1970s

8. Clifford F. Eve to Glenn H. Brown, 21 February 1966, personnel files of the Liquid Crystal Institute, Kent State University Archives and Special Collections, Kent State University Library, Kent, Ohio.

9. "The Mesoscope," *IRIS Proceedings*, Vol. 5, 1960.

10. J.L. Fergason, 1966. U.S. patent 3,529,156, filed June 13, 1966, and issued Sept. 15, 1970.

11. Linda Hamilton, "Liquid Crystals," *Invention and Technology,* Spring, 2002, 23.

12. Merck KGaA, Corporate Communications: "100 years of Liquid Crystals at Merck: The history of the future." March 2004

13. Gary L. Waterman, Wayne E. Woodmansee, "Nondestructive Testing Method Using Liquid Crystals," U.S. Patent 3439525 A, filed Dec. 28, 1966, and issued Apr 22, 1969

14. Theodore Hodson et al., "Encapsulated Cholesteric Liquid Crystal Display Device," 1969. U.S. Patent 3585381, filed Apr 14, 1969, and issued June 15, 1971.

15. Max Garbuny to Glenn Brown, 17 February 1965, personnel files of the Liquid Crystal Institute, Kent State University Archives and Special Collections, Kent State University Library, Kent, Ohio.

16. "Dr. Brown Given Leave as Dean for Research," *The Kent State University Summer News*, Volume XIV, No. 9, August 22, 1968.

17. Glenn H. Brown, James L. Fergason, "First Annual Report of the Liquid Crystal Institute," 6-7. Archives of the Liquid Crystal Institute, Kent State University Archives and Special Collections, Kent State University Libraries, Kent, Ohio.

18. "Two Liquid Crystal Phases with Nematic Morphology in Laterally Substituted Phenylenediamine Derivatives." *Molecular Crystals and Liquid Crystals* 8, (1970): 577.

19. Letter labeled "personal and confidential" from Robert I. White to Glenn Brown, Richard Dunn, James Fergason and Daniel Jones. 26 August 1968, KSU Archives and Special Collections.

20. Glenn H. Brown to Robert I. White, Interdepartmental Correspondence, 30 August 1968, Glenn H. Brown Papers, KSU Archives and Special Collections.

21. Glenn Brown to Robert White, 19 September 1968, Glenn H. Brown Papers, KSU Archives and Special Collections.

22. Elizabeth Popp Berman, *Creating the Market University: How Academic Science Became an Economic Engine* (Princeton: Princeton University Press, 2011), 95.

23. J.L. Fergason, "Liquid Crystals in Nondestructive Testing," *Applied Optics* (1968): 1729-1737

24. J.L. Fergason and T.B. Harsch, 1972. "Thickness Measuring Method and Apparatus," U.S. patent 3,795,133, filed February 15, 1972, and issued March 5, 1974.

25. David N. Kaye, "Combat Coverage of Liquid Crystals," *Electronic Design News,* 1970.

26. Bob Johnston, *We Were Burning: Japanese Entrepreneurs and the Forging of the Electronic Age,* (New York: Basic Books, 1999), 104.

27. M. Schadt and W. Helfrich, "Voltage-Dependent Optical Activity of a Twisted Nematic Liquid Crystal," *Applied Physics Letters* 18, no. 4 (1971): 127-128.

28. Joseph A. Castellano, *Liquid Gold,* (New Jersey: World Scientific Publishing, 2006), 72.

29. Benjamin Gross. "Crystallizing Innovation: The Emergence of the LCD at RCA, 1951-1976," PhD diss., Princeton University, 2011. Gross wrote that, "RCA's insistence that members of its technical staff document their research to protect against future intellectual property litigation ensured the accumulation of extensive notebook holdings, patent records, and internal correspondence related to liquid crystal display (LCD) research." 115-116.

30. Martin Schadt, pg. 13, http://www.eurasc.org/Galeries/AG_2010/schadt/schadt.swf

31. James L. Fergason and Duane E. Werth, 1973. "Liquid Crystal Display Assembly," U.S. Patent 3963324, filed Sept. 10, 1973, and issued June 15, 1976.

32. J.L. Fergason, 1973. "Gasket for Liquid Crystal Light Shutters," U.S. Patent 3853392, filed Sept. 13, 1973, and issued Dec. 10, 1974.

33. Thomas W. Gerdel, "Timely Crystals Spell Jobs, Profits," *The Plain Dealer*, Tuesday, June 19, 1973.

34. J.L. Fergason, 1971. "Display Devices Utilizing Liquid Crystal Light Modulation," U.S. patent 3,731,986, filed Apr. 22, 1971, and issued May 8, 1973.

35. J. L. Fergason, 1973. "Liquid-crystal non-linear light modulators using electric and magnetic fields," U.S. Patent 3918796, filed Jan. 24, 1973, and issued Nov. 11, 1975; J. L. Fergason and T. B. Harsch, 1973. "Reflective system for liquid crystal displays," U.S. Patent 3881809, filed May 25, 1973, and issued May 6, 1975; J. L. Fergason and D. E. Werth, 1973. "Liquid crystal display assembly," U.S. Patent 3963324, filed Sept. 10, 1973, and issued June 15, 1976; J. L. Fergason, 1973. "Gasket for liquid crystal light shutters," U.S. Patent 3853392, filed Sept. 13, 1973, and issued Dec. 10, 1974.

36. Hirohisa Kawamoto, "The History of Liquid Crystal Displays," *Proceedings of the IEEE* (April, 2002), 460-500.

37. Martin Schadt, "Contributions to Today's Liquid Crystal Display Technology," European Academy of Sciences, http://www.eurasc.org/Galeries/AG_2010/schadt/schadt.swf, 4.

38. Daniel C. Jones to Bernard Hall, 25 Jan. 1971, KSU archives.

39. "Former Prof sues KSU, Timex Inc. for $60 million," Sun Newspaper (Cleveland, OH), December 4, 1974.

40. The original trial documents are at the National Archives and Records Administration-Great Lakes Region, 7358 South

Pulaski Rd, Chicago, IL 60629-5898, Ph: (773) 948-9001; Fax: (773) 948-9050; E-mail: chicago.archives@nara.gov; http://www.archives.gov/great-lakes/Case Number: 74CV1087, Accession Number: 021-87-0179, Box Numbers: 7-11 (5 cubic feet) Location: 344/368053-368057

41. R. Lawrence Dessem, "A Government of Laws and Also of Men: Judge William K. Thomas," *Ohio State Law Journal*, Vol. 62, No. 4 (2001): 1327.

42. C. Von Planta, Physics Today, April, 1983, p. 90.

43. T. Sluckin, D. Dunmur and H. Stegemeyer, *Crystals That Flow,* (Taylor and Francis, 2004), 475.

44. Gerhard H. Buntz, "Twisted Nematic Liquid Crystal Displays (TN-LCDs), an Invention from Basel with Global Effects," Information No. 118, Oct. 2005, Internationale Treuhand AG, Basel, Genf, Zurich (in German). In English at http://www.lcd-experts.net/.

45. "Kent State Finds Liquid Crystals Help in Detection of Breast Cancer," *The Plain Dealer,* July 28, 1974

46. United States Environmental Protection Agency, "A Look at EPA Accomplishments: Twenty-five Years of Protecting Public Health and the Environment," http://www2.epa.gov/aboutepa/look-epa-accomplishments-25-years-protecting-public-health-and-environment

47. Jim testified before the 104th and 105th Congresses on Title II, "Early Publication of Patent Applications" and Title IV, "Inventor Protection of the Moorhead-Schroeder Patent

Reform Act. See: http://www.gpo.gov/fdsys/pkg/CRPT-104hrpt879/html/CRPT-104hrpt879.htm

48. T. Sluckin, D. Dunmur and H. Stegemeyer, *Crystals that Flow,* Taylor and Francis, 2004, page 475.

49. Karen O'Leary, "Masterminds: Interviews with Local Inventors, *Gentry,* April, 1999, 87.

RESOURCES

ORIGINAL MANUSCRIPT

James L. Fergason as told to Arthur L. Berman, *Birth of a Technology: A Biographical History of the Liquid Crystal Display,* unpublished manuscript, 2008.

PAPERS BY JAMES FERGASON

James L. Fergason, "Liquid Crystals," *Scientific American,* August, 1964

J. L. Fergason, J.R. Hanson and A. Okaya, "Display of Infrared Laser Patterns by a Liquid Crystal Viewer," *Applied Optics,* 3, 8 (1964).

John T. Crissey, J.L. Fergason, Jack M. Bettenhausen, "Cutaneous Thermography with Liquid Crystals," Presentation, Twenty-sixth Annual Meeting of The Society for Investigative Dermatology, Inc., New York, N.Y., June 20, 1965.

J.L. Fergason, "Liquid Crystals in Nondestructive Testing," *Applied Optics* (1968): 1729-1737

J.L. Fergason and G.H. Brown, "Liquid Crystals in Living Systems," *The Journal of the American Oil Chemists' Society* 45, no. 3, (1968): 120-127.

N. Bravo, J.W. Doane, S.L. Arora, J.L. Fergason, "NMR Study of Molecular Structure in a Fluorinated Liquid Crystal Schiff Base," *The Journal of Chemical Physics* 50, no. 3, (1969): 1398.

Fergason, Taylor and Harsch, "Liquid Crystals and Their Applications," *Electro Technology*, Jan. 1970, 41-50.

J. L. Fergason, "Liquid Crystals and Science," *Aldrichimica Acta*, Vol. 3, No. 3, 1970.

A. Can, M. Chapman, T.W. Davison, K.L. Ewing, J.L. Fergason, C.C. Voorhis, "Detection of Breast Cancer by Liquid Crystal Thermography, A Preliminary Report," *Cancer* 29, no. 5, (1972): 1123-32.

T.W. Davison, K.L. Ewing, J.L. Fergason, "Effects of Activity, Alcohol, Smoking, and the Menstrual Cycle on Liquid Crystal Breast Thermography," *The Ohio Journal of Science* 73, no. 1 (1973): 55-58.

T.W. Davison, K.L. Ewing, J.L. Fergason, N.P. Mulla, N. Sayat, "Liquid Crystal Thermographic Placental Location," *Obstetrics and Gynecology* 42, no. 4 (1973): 574-80.

James L. Fergason, "Performance of a Matrix Display Using Surface Mode," 1980 Biennial Display Research Conference, 1980, IEEE, pg. 177.

James L. Fergason, James A. McCoy. "A New Imaging Paradigm for Medical Applications," Medicine Meets Virtual

Reality, in Studies in Health Technology and Informatics, Vol. 50, Westwood et al., eds., (1997): 278-283.

SPEECHES

James L. Fergason, Induction Speech (National Inventors Hall of Fame, Akron, Ohio), Sept. 19, 1998.

PATENTS BY JAMES FERGASON CITED IN BOOK

J.L. Fergason, T.P. Vogl and M. Garbuny. 1960. "Thermal Imaging Devices Utilizing a Cholesteric Liquid Crystalline Phase Material," U.S. patent 3,114,836, filed March 4, 1960, and issued December 17, 1963.

J.L. Fergason, 1963. "Analytical Method and Devices Employing Cholesteric Liquid Crystalline Materials," U.S. patent 3,409,404, filed Nov. 13, 1963, and issued November 5, 1968.

J.L. Fergason, 1971. "Display Devices Utilizing Liquid Crystal Light Modulation," U.S. patent 3,731,986, filed Apr. 22, 1971, and issued May 8, 1973.

J. L. Fergason, 1973. "Gasket for liquid crystal light shutters," U.S. Patent 3853392, filed Sept. 13, 1973, and issued Dec. 10, 1974.

J. L. Fergason and T. B. Harsch, 1973. "Reflective system for liquid crystal displays," U.S. Patent 3881809, filed May 25, 1973, and issued May 6, 1975.

J.L. Fergason, 1973. "Liquid-crystal non-linear light modulators using electric and magnetic fields," U.S. Patent 3918796, filed Jan. 24, 1973, and issued Nov. 11, 1975.

J. L. Fergason and D. E. Werth, 1973. "Liquid crystal display assembly," U.S. Patent 3963324, filed Sept. 10, 1973, and issued June 15, 1976.

PRIZES

Lemelson-MIT, "Liquid Crystal Display Trailblazer Receives Largest Prize for Invention in the United States: James Fergason's Achievements Recognized With $500,000 Lemelson-MIT Prize," http://web.mit.edu/invent/n-pressreleases/n-press-06LMP.html

Finalist, 1992 Discover Awards: Sight," Oct. 1992, http://discovermagazine.com/1992/oct/1992discoverawar134#.UqucSSf9y1h

VIDEOS

Inventing the Future for Fun and Profit, 1993; Woody Clark Productions, Center for New Venture Alliance, School of Business & Economics, California State University in cooperation with U.S. Department of Energy, VHS.

"Liquid Crystals," Interface science show, VHS, Cal Tech, 1970.

"Fergason on Patent Reform," posted by Symax, YouTube video, 3:57, Jan. 3, 2008, http://www.youtube.com/watch?v=c7h6Z4MlqIc

OTHER RESOURCES

Sardari L. Arora, 1973. "Field Effect Light Shutter Employing Low Temperature Nematic Liquid Crystals," U.S. Patent 4086002, filed Nov. 20, 1973, and issued Apr. 25, 1978.

Elizabeth Popp Berman, *Creating the Market University: How Academic Science Became an Economic Engine* (Princeton: Princeton University Press, 2011).

Joseph A. Castellano, *Liquid Gold,* (New Jersey: World Scientific Publishing, 2006).

Benjamin Gross, "How RCA Lost the LCD," IEEE Spectrum, Nov. 1, 2012.

Linda Hamilton, "Liquid Crystals," *Invention and Technology,* Spring 2002

Thomas Harsch, "The Twisted Nematic Discovery Narrative," August, 2013, 3.

Bob Johnston, *We Were Burning: Japanese Entrepreneurs and the Forging of the Electronic Age* (New York: Basic Books, 1999).

Hirohisa Kawamoto, "The History of Liquid Crystal Displays," *Proceedings of the IEEE* (April, 2002), 460-500.

David N. Kaye, "Combat Coverage of Liquid Crystals," *Electronic Design News,* 1970.

O. Lehmann, "On Flowing Crystals," Zeitschrift fur Physikalische Chemie 4, 462-472, 1889.

T. Martin Lowry, *Optical Rotatory Power,* Longsman, Green & Co., London, 1935. U.S. edition published by Dover in 1964.

Michael S. Malone, "The Smother of Invention," Forbes.com, June 24, 2002, http://www.forbes.com/asap/2002/0624/032_2.html

Official Website of the Nobel Prize, "History and Properties of Liquid Crystals," http://www.nobelprize.org/educational/physics/liquid_crystals/history/index.html

J.G. Pritchard, *Poly (Vinyl Alcohol): Basic Properties and Uses*, (McDonald Technical & Scientific: 1970).

E.P. Raynes, "Twisted Nematic Liquid-Crystal Electro-optic Devices with Areas of Reverse Twist," *Electronic Letters*, March 8, 1973, Vol. 9, No. 5.

E.P. Raynes, "Improved Contrast Uniformity in Twisted Nematic Liquid Crystal Electro-optic Display Devices," 10, *Electronic Letters* (1974): 141-142

M. Schadt and W. Helfrich, "Voltage-Dependent Optical Activity of a Twisted Nematic Liquid Crystal," *Applied Physics Letters* 18, no. 4 (1971): 127-128.

Martin Schadt, "Contributions to Today's Liquid Crystal Display Technology," Journal European Academy of Sciences, NR 1, 2011

Amelia Carolina Sparavigna, "James Fergason, a Pioneer in Advancing of Liquid Crystal Technology," http://arxiv.org/ftp/arxiv/papers/1310/1310.7569.pdf, published Oct. 28, 2013, Cornell University Library, p. 1.

H. Stegemeyer, "Centenary of the Discovery of Liquid Crystals," *Liquid Crystals*, Vol. 5, no. 1, 5-6, 1989

"Optel's Misadventures in Liquid Crystals," *Fortune* (1973): 206.

ACKNOWLEDGMENTS

BEFORE HIS DEATH in 2008, Jim completed a book manuscript titled, *Birth of a Technology: A Biographical History of the Liquid Crystal Display*. We are indebted first to Arthur L. Berman, who helped Jim write this account, answered our questions and provided us with archival materials.

We are also grateful for the extensive assistance of Thomas Harsch for bringing the history of the twisted nematic display, and its inventor, to life and for his constant support and encouragement in ways great and small. His affection for Jim Fergason, his long-time colleague and friend, kept us inspired throughout the writing of this book. In addition, the seven detailed written accounts he gave us of Ilixco's invention of the liquid crystal display and all of its components form the core of the relevant chapters on Ilixco. His recollections were also instrumental in helping us construct the chapter on the American Xtal Chemical Corporation (ALX). Tom also provided numerous photographs and illustrations that help describe the physical properties of liquid crystals.

We are very grateful to chemist Kenneth Marshall for the hours he spent describing the chemical work he did at Ilixco and ALX and patiently responding to our many questions. He also fact-checked portions of the manuscript and entertained us with stories about explosions in the lab.

Dora Fergason, Jeff Fergason, John Fergason and Susan Fergason provided a wealth of information on Jim's life and inventions. Dora, Jeff and John also helped with the final fact check of the book.

We would like to thank Tom Davison for his account of working with Jim on the biological applications of liquid crystals.

Frederick Kahn shared memories of seeing the early Ilixco displays.

Carole Garbuny Vogel gave us information on her father, Max Garbuny.

David N. Kaye gave us his personal account of interviewing Jim on the Kent State campus in Ohio on May 4, 1970.

Our thanks also go to Duane Werth, Karen MacDonald, Blake Brown, Beth Cunningham and Sharen Breyer for their anecdotes about working with Jim.

Zachary Paris and R. Lawrence Dessem were law clerks for the late Judge William K. Thomas, who presided over an important patent case Jim fought during the 1970s. They gave us valuable insights into the judge's working style.

Warren Sklar, Jim's friend and patent lawyer for many of his later patents, lent us valuable materials, answered many questions and demonstrated the NCAP invention for Marian.

Charles (Chuck) McLaughlin filled in the history of Jim's work at the Taliq Corporation.

We would also like to thank Thomas Shunk, the lead counsel for Jim in the patent infringement suit against Tektronix and Jay Campbell, who assisted him, for discussing the ins and outs of the case.

Herbert Wamsley, the executive director of the Intellectual Property Owner's Association (IPO), discussed Jim's work as a board member of the IPO and his testimony before Congress.

Nicholas P. Godici graciously emailed us an account of Jim's service on the Patent Public Advisory Committee.

Jim's grand-niece Emily Cantrell verified genealogical information and relayed stories about Jim's childhood from her interviews with Jim's brother Lewis. Emily also interviewed Jim's niece, Carol Jean Bradley, for information about Jim's personality as a child.

We're grateful to Hermon Joyner and Melissa Schaaf for their extensive feedback on an early draft of this manuscript.

We extend special thanks go to Amanda Faehnel, public services librarian, special collections and archives, Kent State University library where we were able to find original correspondence, documents, and memos at the time that Jim was at LCI.

We would like to thank Over and Above Creative that helped make a manuscript into a book. A big thank you to Rick Benzel for his editing, rewriting and support, Susan Shankin for her design work on the cover, and the layout of the book, and Tim Kummerow who prepared the illustrations. And finally, Daniel Baxter who created a wonderful illustration for the cover of the book.

Finally, Terri would like to thank her husband Randy for his patience and support through this extensive process.

ABOUT THE AUTHORS

 TERRI FERGASON NEAL is the daughter of James L. Fergason. Science and art are her passions. She earned her first degree in Biology at Hiram College, then a Masters in Zoology at Arizona State. Her first experiences working in science came from working as a lab tech, then chemist, in one of her father's early ventures. She continued working in research at Barrow's Neurological Institute in Phoenix, AZ. She now lives as an aspiring artist in the Pacific Northwest. Terri is proud to complete Jim's story and his views and recollections of the early days of the liquid crystals industry.

MARIAN PIERCE is an editor, researcher and writer in Portland, OR.

Visit mrliquidcrystal.com to see color photos related to this book.

---

*If you enjoyed reading this book, please share it with others on Facebook.
We'd also appreciate your review of it on Amazon and on other book review websites.*

CPSIA information can be obtained
at www.ICGtesting.com
Printed in the USA
LVOW13s1005261216
518699LV00011B/1014/P